A. A. Gershon
C. H. Calisher
A. M. Arvin (eds.)

**Immunity to
and Prevention of
Herpes Zoster**

Springer-Verlag Wien GmbH

Prof. Dr. Anne A. Gershon
Department of Pediatrics, College of Physicians & Surgeons,
Columbia University, New York, USA

Prof. Dr. Ann M. Arvin
Department of Microbiology and Immunology, Stanford University,
Stanford, USA

Prof. Dr. Charles H. Calisher
Arthropod-borne and Infectious Diseases Laboratory,
Colorado State University, Fort Collins, USA

© 2001 Springer-Verlag Wien
Originally published by Springer-Verlag Wien New York in 2001
Softcover reprint of the hardcover 1st edition 2001
Typesetting: Thomson Press (India) Ltd., New Delhi

Cover image: An electron micrograph of a biopsy obtained from the skin of a patient during an episode of acute zoster. The biopsy has passed through a vesicle in an affected region of the epidermis. The extracellular space has expanded and deeply indents the surface of a keratinocyte. The vesicle fluid contains fully enveloped varicella zoster virions. Notice that many of the viral particles are well formed and intact. An enveloped viral particle that has not yet been released by exocytosis can also be seen within the keratinocyte at the lower left of the field. Notice that this viral particle is also well formed and is enclosed by itself within a small transport vacuole. Original magnification: × 54,000

Courtesy of Prof. Michael Gershon

Printed on acid-free and chlorine-free bleached paper

SPIN: 19780115

With 42 Figures

CIP data applied for

ISBN 978-3-211-83555-5   ISBN 978-3-7091-6259-0 (eBook)
DOI 10.1007/978-3-7091-6259-0
Archives of Virology [Suppl] 17

Preface

Under sponsorship of the National Institutes of Health of Japan, an international conference entitled "Immunity and Prevention of Herpes Zoster" was held in Osaka, Japan, March 8–10, 1999.[1] Attendees included basic and clinical investigators from Asia, Europe, and North America.

The meeting was organized to explore progress made in basic virology and molecular understanding of varicella zoster (VZV), and to provide information on current knowledge of latency of VZV in humans. Updates on the immunology responses of humans to VZV, and a description of the current status of varicella vaccine worldwide were also included. In addition, the possibility of preventing zoster in people latently infected with wild-type VZV by immunizing them with varicella vaccine was presented. The papers in this volume include written summaries of most of the presentations given at that conference.

Coincidentally but appropriately, the conference marked the twenty-fifth or "silver anniversary" of the first publication of the development and use of live varicella vaccine to prevent varicella, by Takahashi and his colleagues. Because varicella vaccine is the first herpesvirus vaccine licensed in use for humans, it is of special interest to all individuals who studied these pathogens. In view of the interest in developing vaccines against other herpesviruses, there was also a presentation on the current status of vaccines against cytomegaloviruses (CMV) at the conference.

Dr. Michiaki Takahashi, of Osaka University, the developer of the live attenuated varicella vaccine, introduced the program. Initially thought to be useful mainly to protect against chickenpox, it is hoped that the vaccine will be useful in preventing zoster as well. Definitive data regarding this possibility should be available within the next few years.

For those wishing to place VZV in perspective, it is classified as an alpha herpesvirus, closely related to two other alphaherpesviruses – herpes simplex virus (HSV) 1 and 2. Other human herpesviruses include: CMV, Epstein Barr virus (EBV), and herpesviruses 6, 7, and 8. For those interested in the source of the name chickenpox, it remains uncertain. Theories include that 1. "chickenpox" is a diminutive of smallpox; 2. it is derived from the chick pea, which the skin lesions are said to resemble, and/or 3. that the name is somehow related to the

[1]This meeting was supported by an educational grant from the Research Foundation for Microbial Diseases of Osaka University.

barnyard fowl. Herpes is derived from the Greek word for "creep", and "zoster" is Greek and Latin for "belt". "Shingles" is related to the Latin word for "girdle".

We expect that the articles included in this volume will provide more definitive information on the nature of the virus itself; the nature of the names attached to diseases caused by VZV must forever remain somewhat mysterious.

Anne A. Gershon
Ann M. Arvin
Charles H. Calisher

Contents

Listed in Current Contents

The current status of live attenuated varicella vaccine

A. A. Gershon

Department of Pediatrics, Columbia University College of Physicians & Surgeons,
New York, New York, U.S.A

Summary. This manuscript reviews the means by which live attenuated varicella vaccine offers protection against varicella and zoster. It is accepted that although varicella is usually a mild illness, complications leading to morbidity and mortality are significant and the disease is worth preventing. The vaccine offers close to 100% protection from severe chickenpox and 90% protection from illness. Waning of immunity after vaccination, particularly in children, has not been a significant problem. Ways in which vaccination may decrease the incidence and severity of zoster include the following. Vaccine virus may be less likely to establish latency and to be able to reactivate than wild type virus. In addition, by selective immunization of certain hosts such as HIV-infected children whose immune systems are still relatively intact and individuals with latency due to wild type virus to boost the cell-mediated immune response to the virus, zoster may be decreased. Varicella vaccine is predicted to have a major impact on the epidemiology of varicella and zoster in countries with high vaccine uptake.

Introduction

During the past 25 years much has been learned about the safety and efficacy of varicella vaccine in both healthy and immunocompromised patients for prevention of varicella (chickenpox). At present this is the only vaccine against a herpesvirus for use in humans that has achieved licensure in any country. The Oka strain of virus is now licensed in many countries worldwide. This presentation will review salient points about varicella vaccine and will also emphasize various mechanisms by which vaccination might prevent not only varicella but zoster as well.

The ideal response to varicella vaccine

The course of a physician who was the first susceptible adult to be immunized against varicella in the United States, 20 years ago, serves to illustrate what might be considered an optimal response to varicella vaccine. Prior to immunization, at age 30, he had no history of varicella or detectable humoral or cellular immune responses to varicella zoster virus (VZV). Ten days after he was vaccinated, he developed a small papule on his neck from which VZV could not be cultured. At about the same time, he developed specific cell-mediated immunity (CMI) to VZV measured by lymphocyte stimulation, and antibodies measured by the fluorescent antibody to membrane antigen (FAMA) assay. After several months, his immune responses to VZV decreased, but not to their original pre-immunization levels. Almost 4 months later, he was closely exposed to a child with varicella in the emergency room at Bellevue Hospital in New York City. This child had severe varicella with respiratory failure and this physician administered mouth-to-mouth resuscitation to the child. Following this exposure, the physician-vaccinee developed prominent boosting of his humoral and CMI to VZV by a factor of about 10, but no symptoms of varicella. Seven years after vaccination, both his children developed clinical varicella, within 2 weeks of each other. Despite this prolonged and intimate exposure to VZV, he did not become ill, and he had no further boosting of his immunity to the virus. He has not developed varicella in the now 20 years since he was immunized. He continues to have antibodies to VZV detectable by several assays, including FAMA and a commercially available enzyme-linked immuno-assay (ELISA) test that measures IgG to VZV.

Current problems associated with vaccine use in the United States

While the above representative example illustrates the potential utility of varicella vaccine, many physicians in the United States remain wary of this vaccine. There seem to be two major misconceptions involved in this view. One is that varicella is a minor illness not worthy of prevention by vaccination. The second is that following immunization, there is significant waning of immunity [8]. Both of these misconceptions can be refuted.

To refute the first, there is abundant evidence that varicella can be a severe illness even in persons who were healthy before developing chickenpox. Every year about 100 individuals die of varicella in the United States, half of them children. Complications associated with significant morbidity and mortality include severe bacterial superinfections such as those caused by streptococci and staphylococci localized to soft tissues, bones, and lungs. Encephalitis, while uncommon, is often severe or fatal. Adults are at increased risk from these complications of varicella and may develop primary varicella pneumonia. It is not possible to predict in advance whether healthy individuals will develop severe illness, highlighting the importance of prevention. Finally, varicella is likely to be severe or fatal in immunocompromised patients. Having immunity to VZV prior to becoming immunocompromised is an obvious advantage.

With regard to the second misconception, there is little evidence that immunity to varicella wanes after immunization, particularly in children. One approach to assess waning immunity is to examine VZV antibodies to determine whether they decrease significantly with time. In studies of children from Japan and the United States, involving over 500 vaccinees for as long as 20 years, no significant loss of antibodies to VZV has been detected [1, 3, 15]. In some studies, the antibody titers to VZV in immunized children have actually increased with time, presumably due to boosting from re-esposure to VZV or even sublclinical reactivation of latent virus [2, 3, 10]. Development of varicella is known to occur in about 10% of vaccinated children; however these illnesses, referred to as breakthrough varicella, are a modified form of illness.

The situation is somewhat more complicated in adult vaccinees, because some 20% of vaccinated adults may lose detectable VZV FAMA antibodies with time. However, their breakthrough illnesses have been reported as modified and an increase in the frequency of breakthrough illness with time has not been recorded [10, 13].

Thus another, perhaps more informative method to assess the possibility of waning immunity to VZV is to determine whether there is any increase in the number and severity of cases of varicella with time after immunization. As just mentioned, this has not been observed in adult vaccinees. In addition, in both leukemic [10] and healthy children [15] who were vaccinated, there has been no indication of an increase in either incidence or severity of breakthrough chickenpox with time. Therefore waning immunity has not been demonstrated to be a significant problem for varicella vaccinees.

Post-licensure studies in the United States mandated by the Food and Drug Administration (FDA) are continuing to examine VZV antibody titers year after immunization and also the incidence and severity of varicella in vaccinees with time. A case control study of the protective efficacy of varicella vaccine in clinical practice has indicated that the vaccine is about 85% protective in healthy immunized children [18]. Most of those with breakthrough varicella have experienced a mild illness.

The effect of varicella vaccine on zoster

Four independent studies have indicated that, in immunocompromised children, the incidence of zoster is lower after vaccination than after natural infection. In two studies, one from Japan and one from the United States, in which historical

Table 1. Events following immunization of a 30 year old physician against varicella in 1979

Event	Comments
No history of varicella Immunization (1 dose)	No detectable antibodies or CMI to VZV
10 days later	Papule on neck Detectable antibodies and CMI to VZV
Next 3 months	Decrease in levels of antibodies and CMI to VZV
Four months after immunization	Intimate exposure to patient in emergency room, *In extremis* from varicella
Within a week	Increase (10×) in antibodies and CMI to VZV
Two to three weeks later	No evidence of varicella
Seven years later	Household exposure to two children with varicella
Two to three weeks later	No evidence of disease; no boost in immunity

controls were used, the incidence of zoster in the vaccinated was extremely low (0% to 6%), compared to controls (19–21%) [6, 19]. In a study from the United States, in which matched controls were included, the incidence of zoster was 2% in vaccinees and 16% in controls [14]. In a study of French children who were immunized prior to undergoing renal transplantation, the incidence of zoster was 7% in the immunized and 13% in controls [5]. Development of zoster has been related to prior presence of a VZV-associated rash, either due to vaccine or wild type VZV [10]. Presumably, replication of VZV in the skin enables VZV to reach proximal nerve endings and then to establish latent infection in the dorsal root ganglia. Another possibility is that without rash there may be less latent infection. It is further possible that the attenuated Oka strain is less likely to be able to reactivate than wild type VZV. Hypothetically, both of these factors could be operative to decrease the incidence of development of zoster in vaccinated persons.

The relationship between CD4 levels at onset of primary infection (varicella) and subsequent clinical reactivation

Longitudinal studies of children with underlying infection with human immuno-deficiency virus (HIV) who developed clinical varicella suggest the importance of CD4 lymphocytes in control of VZV. Only 6% of HIV-infected children (n = 32) who developed varicella when they had more than 15% CD4 cells subsequently developed zoster over a 2 year interval, while 80% (n = 13) developed zoster within 2 years if they had < 15% CD4 cells when they contracted varicella [9, 12]. Based on this experience, it was decided to immunize HIV-infected children who had more than 25% CD4 cells against varicella. It was hoped that by providing primary immunity to VZV in the setting of relatively normal CD4 levels, that the incidence of zoster in these children would decrease. This study was carried out by the Pediatric AIDS Clinical Trials Group (ACTG) in 1998. In this study, 41 HIV-infected children with more than 25% CD4 cells were vaccinated against chickenpox. These children tolerated varicella vaccine much as did healthy children. Five percent developed a mild vaccine-associated rash, and an immune response to VZV was detected in 88%. One of these children developed a mild case of varicella following exposure to VZV, and one developed what was diagnosed as clinical zoster but which could not be documented virologically [16]. These children are continuing to be followed for the incidence of varicella and zoster. Thus far, varicella vaccine appears to have been beneficial to them.

Prevention or modification of zoster in high-risk individuals by immunization

There has been considerable interest in this possibility since 1984, when it was shown that CMI responses to VZV could be boosted by immunization with varicella vaccine [4]. Since that time, additional studies have shown that by boosting of CMI to VZV either by immunization or through natural exposure to the virus that the rate and severity of zoster declines [11, 17]. A double-blind, randomized, placebo-controlled study of varicella vaccination in healthy persons over 60 years old is currently being carried out in the United States.

Current routine use of varicella vaccine in the United States

Varicella vaccine was licensed for use in the United States in March 1995. At first, routine use of vaccine for children was controversial. Today, however, the vaccine is recommended for all healthy children at one year of age by all of the agencies that make recommendations for vaccine use in children. More and more states are requiring immunization with varicella vaccine before they will allow children to attend day care or school. The CDC aims that by the year 2010 > 90% of children aged 19–35 months will be immunized with > 95% of children immunized by the time they begin to attend school [7]. Thus far, it appears that the incidence of zoster is about 20 times lower in vaccinated children than would be expected had they experienced natural varicella.

Conclusions

There are at least 4 ways by which by which use of varicella vaccine might decrease the incidence or zoster. There may be decreased latency and/or decreased reactivation after immunization compared to the situation after natural infection, due to primary infection with attenuated Oka VZV. The tendency towards reactivation may be further decreased by immunization of high-risk individuals such as HIV-infected children when their CD4 cells remain high. Finally, boosting of CMI to VZV may be achieved by immunization of persons with established latency with wild type VZV, which may result in less reactivation and either less frequent or milder episodes of zoster.

References

1. Arvin A, Gershon A (1996) Live attenuated varicella vaccine. Annu Rey Microbiol 50: 59–100
2. Asano Y, Albrecht P, Vujcic LK, Quinnan GV Jr, Kawakami K, Takahashi M (1983) Five-year follow-up study of recipients of live varicella vaccine using enhanced neutralization and fluorescent antibody membrane antigen assays. Pediatrics 72: 291–294
3. Asano Y, Suga S, Yoshikawa T, Kobayashi H, Yazaki T, Shibata M, Tsuzuki K, Ito S (1994) Experience and reason: twenty year follow up of protective immunity of the Oka live varicella vaccine. Pediatrics 94: 524–526
4. Berger R, Luescher D, Just M (1984) Enhancement of varicella-zoster-specific immune responses in the elderly by boosting with varicella vaccine. J Infect Dis 149: 647
5. Broyer M, Tete MT, Guest G, Gagnadoux MF, Rouzioux C (1997) Varicella and zoster in children after kidney trasplantation: long term results of vaccination. Pediatrics 99: 35–39
6. Brunell PA, Taylor-Wiedeman J, Geiser CF, Frierson L, Lydick E (1986) Risk of herpes zoster in children with leukemia: varicella vaccine compared with history of chickenpox. Pediatrics 77: 53–56
7. Centers for Disease Control (1999) Prevention of varicella. Morb Mort Wkly Rep 48: 1–6

8. Centers for Disease Control (1999) Varicella-related deaths. Florida Morb. Mort Wkly Rep 48: 379–381
9. Derryck A, LaRussa P, Steinberg S, Capasso M, Pitt J, Gershon A (1998) Varicella and zoster in children with human immunodeficiency virus infection. Ped Infect Dis J 17: 931–933
10. Gershon A (1995) Varicella-zoster virus: prospects for control. Adv Ped Infect Dis 10: 93–124
11. Gershon A, LaRussa P, Steinberg S, Lo SH, Mervish N, Meier P (1996) The protective effect of immunologic boosting against zoster: an analysis in leukemic children who were vaccinated against chickenpox. J Infect Dis 173: 450–453
12. Gershon A, Mervish N, LaRussa P, Steinberg S, Lo S-H, Hodes D, Fikrig S, Bonagura V, Bakshi S (1997) Varicella-zoster virus infection in children with underlying HIV infection. J Infect Dis 175: 1496–1500
13. Gershon AA, Steinberg S, LaRussa P, Hammerschlag M, Ferrara A, NIAID-Collaborative-Varicella-Vaccine-Study-Group (1988) Immunization of healthy adults with live attenuated varicella vaccine. J Infect Dis 158: 132–137
14. Hardy IB, Gershon A, Steinberg S, LaRussa P, et al. (1991) The incidence of zoster after immunization with live attenuated varicella vaccine. A study in children with leukemia. N Engl J Med 325: 1545–1550
15. Johnson C, Stancin T, Fattlar D, Rome LP, Kumar ML (1997) A long-term prospective study of varicella vaccine in healthy children. Pediatrics 100: 761–766
16. Levin M, Gershon A, Weinberg A, Blanchard S, Wells B, Nowak B, ACTG 265 Protocol T (1999) Administration of varicella vaccine to HIV-infected children. In: Retroviral Conference, Chicago, December 1999
17. Levin M, Murray M, Zerbe G, White CJ, Hayward AR (1994) Immune responses of elderly person 4 years after receiving a live attenuated varicella vaccine. J Infect Dis 170: 522–526
18. Shapiro ED, LaRussa PS, Steinberg S, Gershon A (1998) Protective efficacy of varicella vaccine. In: 36th Annual Meeting, Infectious Diseases Society of America, Denver, CO
19. Takahashi M, Gershon A (1994) Varicella Vaccine. In: Mortimer E, Plotkin S (eds) Vaccines, 2nd ed. Saunders, Philadelphia, pp 387–417
20. White CJ (1996) Clinical trials of varicella vaccine in healthy children. Infect Dis N Am 10: 595–608
21. White CJ (1997) Varicella-zoster virus vaccine. Clin Infect Dis 24: 753–763

Author's address: Dr. A. A. Gershon, Department of Pediatrics, Columbia University College of Physicians & Surgeons, 650 W. 168th Street, New York, NY 10032, U.S.A.

Evidence for frequent reactivation of the Oka varicella vaccine strain in healthy vaccinees

P. R. Krause

Laboratory of DNA Viruses, Division of Viral Products, Center for Biologics Evaluation and Research, Food and Drug Administration, Bethesda, Maryland, U.S.A.

Summary. Serum antibody levels and infection rates were followed for 4 years in 4,631 children immunized with the recently licensed Oka strain varicella vaccine. Anti-VZV titers declined over time in high-responder subjects, but rose in vaccinees with low titers. Among subjects with low anti-VZV titers, the frequency of clinical sequelae and immunological boosting significantly exceeded the 13%/yr rate of exposure to wild type varicella. These findings indicate that the Oka strain of VZV persisted in vivo, and reactivated as serum antibody titers declined after vaccination. This mechanism may improve vaccine-associated long-term immunity.

Introduction

Wild type varicella zoster virus (wtVZV) causes chickenpox, a common childhood illness characterized by fever and a vesicular rash [7]. WtVZV persists in a latent form in the sensory ganglia, and can re-activate to cause herpes zoster [22]. The public health consequences of chickenpox infection are well documented [7]. Although symptoms are frequently mild, wtVZV can lead to serious complications such as bacterial superinfection, pneumonitis, and death [28].

In 1995, the U.S. Food and Drug Administration licensed a varicella vaccine derived from the Oka strain of virus (OkaVZV). More than 10 million American children have since been inoculated with OkaVZV vaccine. Studies of OkaVZV immunization conducted in Japan indicate that vaccination increases resistance to chickenpox infection for > 20 years [2] . Phase III clinical trials conducted in the U.S. showed that chickenpox rates among OkaVZV vaccinees were reduced by 70–90% over 4 years of follow-up [14, 20]. Interpretation of these findings was complicated by the observation that vaccinees were periodically exposed to wtVZV [2, 26], which may have boosted anti-varicella immunity and prolonged vaccine efficacy [12, 14]. This manuscript expands upon a previous publication [15] indicating that asymptomatic reactivation of OkaVZV in vaccinees may also have contributed to these rising antibody titers.

Materials and methods

A computerized database containing antibody titers and chickenpox rates from all children vaccinated with OkaVZV prior to its 1995 U.S. licensure was provided by Merck & Co. (Whitehouse Station, NJ). This included 4,631, 3,219, 1,248, 446 and 456 children at 6 wk, 1, 2, 3 and 4 years after vaccination, respectively. A total of 243 children had titers drawn at all 5 time points. No titers were determined following chickenpox. Antibody titers were measured using a highly sensitive and reproducible gpELISA [13, 25]. The standard deviation of this assay was 0.3 units (U). Titers $\geq 0.6U$ were considered positive. A vaccinee was considered seronegative if his/her antibody titer was $< 0.6U$, and was assigned a titer of 0.6U for purposes of calculation (requiring that all > 4-fold increases (boosts) yield titers ≥ 2.5 U). Year-to-year fold-changes in antibody titer were calculated for each individual. Cumulative results are expressed as geometric mean fold increases $\pm$ standard·error. For comparison of outcomes within groups of children with similar antibody titers, titers were divided into prospectively chosen ranges of $< 1.25U$, 1.25–2.5U, 2.5–5U, 5–10U, 10–20U, 20–40U, and $\geq 40U$, such that 2-fold increases in titer would cause a single-range shift, and > 4-fold increases would cause at least a 2-range shift. Titers at breakpoints between these ranges were assigned to the higher range. All statistical calculations were performed using Microsoft Excel. Chi square analyses were used to compare proportions. The Wilcoxon rank sum test was used to compare fold changes in antibody titer between children with higher vs. children with lower titers. To compare combined infection and boost rates with the background wtVZV exposure rate, the critical ratio (z_c) was calculated, and p values were determined in accordance with the normal approximation to the binomial distribution [8]. For analyses stratified by age, vaccinees were categorized into the 1–4, 5–6, and > 6 yr age brackets. All rates were expressed as annual rates.

Results

To study potential viral reactivation in OkaVZV vaccinees, we analyzed a database containing clinical outcomes and the results of antibody titers determined 6 wk post vaccination and yearly thereafter, from 4,631 children aged 1–13 years who received the vaccine prior to its licensure in the United States. Anti-VZV titers and infection rates of all children participating in the three licensure trials of the OkaVZV vaccine (initiated in 1982, 1987 and 1992) were studied. The average infection rate among vaccinees was $2.7 \pm 0.6\%/yr$ (253/9,544) versus 11.6–13.0%/yr in previously seronegative children in the placebo-control arm of one of the studies [16,27]. Contemporaneous epidemiological studies of chickenpox rates in children of similar ages yielded estimates of wild type VZV exposure of 4.6–9.1%/yr [6, 9, 28]. However, these epidemiological studies included children with previous chickenpox infection, and thus would be expected to underestimate wtVZV exposure rates. Higher antibody titers were uniformly associated with lower infection rates (Table 1). The 0.07%/yr rate of chickenpox among children with the highest titers (≥ 40) indicates that diagnosis of chickenpox was accurate.

Consistent with previous reports, mean antibody levels rose over time at a geometric mean rate of 1.66 fold/yr in subjects with titers less than 10U (Table 1, line D). However, they declined at a geometric mean rate of 0.75 fold/yr in subjects with anti-VZV titers $\geq 10U$. The distributions of annual fold-changes in anti-

Table 1. Effect of OkaVZV vaccination on infection and immunologic boosting

Line	Analysis	Antibody titer range (in units)							
		< 1.25	1.25–2.5	2.5–5	5–10	10–20	20–40	> 40	Total
	All subjects								
A	# Evaluated	540	651	1,197	1,947	2,573	1,250	1,386	9,544
B	# Infected[a]	66	53	51	46	27	9	1	253
C	Infection rate (%)	12.2[b]	8.1[b]	4.3[b]	2.4[b]	1.0	0.7[b]	0.07	2.7
D	Avg fold titer increase	4.75	2.30	1.5	1.44	1.05	0.90	0.33	1.07
	Titer pairs evaluated for annual change								
E	# Evaluated	202	304	592	995	1,353	583	688	4,717
F	# with 4-fold boost	46	34	89	136	133	55	15	508
G	Avg fold titer increase	63.4	55.9	33.9	25.0	19.1	14.1	8.1	25.4
H	# Infected[a]	2	1	1	0	0	0	0	4
I	Infection rate (%)	4.3	2.9	1.1	0	0	0	0	0.8
J	# with no titer increase	59	155	276	455	757	387	599	1931
K	# Infected[a]	32	26	19	17	8	1	0	103
L	Infection rate (%)	54.2	16.8	6.9	3.7	1.1	0.3	0	5.3

Subjects were stratified by antibody titer (expressed in units, see text; obtained 6 weeks to 4 years after immunization). Data include the number and percent of subjects with chickenpox, with > 4-fold increases in anti-VZV titer within 12 months of the initial titer determination, or with no titer increase from one year to the next. Data are expressed as annual rates

[a]Infection is defined as clinically evident chickenpox

[b]Statistically different ($p < 0.05$, Chi Square) from next higher titer range

VZV titers differed significantly between these groups ($p < 0.001$, Wilcoxon test), indicating that subjects with low titers were more likely to experience year-to-year increases in titer.

To confirm this observation, anti-VZV titers were independently studied in each subject at each time point after immunization. To identify exposures to VZV, a boost was defined as a > 4-fold increase in anti-VZV titer from one year to the next. Clinicians and other investigators routinely accept a 4-fold increase in anti-VZV titers as evidence of VZV re-exposure [1, 3, 18, 23, 24, 27]. Moreover, a > 4-fold rise in titer correlated with significantly improved protection from infection (Table 1, line C) and thus represents a clinically meaningful endpoint. Higher titers also were associated with lower infection rates in each individual year of follow-up (Table 2, Spearman's rank correlation r_s range: -0.96 to -1.0). Indeed, in cases of intentional re-vaccination or household exposure to wtVZV, only about half of vaccinees (including subjects with low anti-VZV titers) developed a > 4-fold increase in titer. Based on these data, we conclude that although every exposure to VZV may not elicit a > 4-fold boost in titer, the presence of a > 4-fold titer boost provides evidence of a significant VZV exposure.

Table 2. Infection rate (% per year) ±SE by titer (in units, see text) and year after administration of OkaVZV vaccine

Titer (U)	Infection rate (%/yr)				
	All year	year 1	year 2	year 3	year 4
< 1.25	12.2 ± 1.4	10.5 ± 2.0	12.4 ± 2.3	16.7 ± 4.4	13.5 ± 5.6
1.25–2.5	8.1 ± 1.1	9.3 ± 1.7	6.6 ± 1.6	9.0 ± 3.0	8.1 ± 4.5
2.5–5	4.3 ± 0.6	3.6 ± 0.8	4.0 ± 0.9	6.5 ± 2.0	6.5 ± 3.1
5–10	2.4 ± 0.3	1.7 ± 0.4	2.6 ± 0.6	3.8 ± 1.2	5.5 ± 2.7
10–20	1.0 ± 0.2	0.9 ± 0.2	1.3 ± 0.4	0.6 ± 0.5	3.1 ± 1.8
20–40	0.7 ± 0.2	0.7 ± 0.3	0.7 ± 0.4	0	3.8 ± 2.6
≥ 40	0.07 ± 0.07	0	0	0.4 ± 0.4	0

Among the 4,717 consecutive pairs of yearly antibody titers determined, there were 508 boosts (Table 1, lines E,F). The combined rate of immunologic boosting plus clinical chickenpox infection was stratified by antibody titer. In vaccinees with titers ≥ 10U, the infection rate was 0.7%/yr while the combined infection plus boost rate was 8.4 ± 0.6%/yr (Fig. 1). In contrast, the combined infection plus boost rate in subjects with titers < 10U was 18.8 ± 1.1%/yr (2.2-fold higher than those with titers ≥ 10U, $\chi^2 = 123$, $p < 0.001$). In subjects with titers < 2.5U the infection plus boost rate was 24.2 ± 2.4%/yr (2.9-fold higher, $\chi^2 = 76$, $p < 0.001$), and in subjects with titers < 1.25, the infection plus boost rate was 32.2 ± 4.2%/yr (3.8-fold higher, $\chi^2 = 139$, $p < 0.001$). These data also demonstrate that the combined infection plus boost rate among children with antibody titers below 10 U significantly exceeded the wild-type VZV exposure rate of 13%/yr ($z_c = 7.9$, $p < 0.001$) [16, 27].

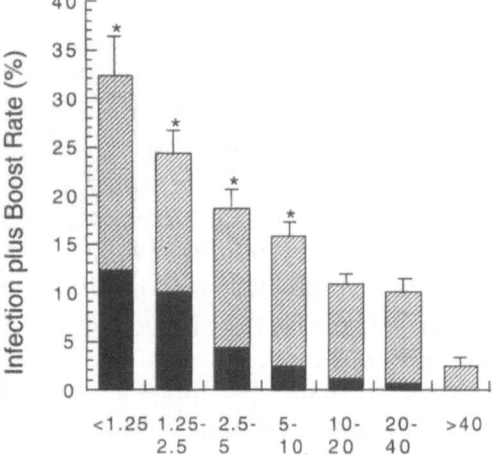

Fig. 1. Antibody titers were obtained on initially seronegative vaccinees 6 wk post immunization and yearly thereafter. The combined rate of varicella infection (solid bars) plus > 4-fold titer increases (boosts, crosshatched bars) during the subsequent 12 months is shown. An asterisk denotes that the combined infection and boost rate significantly ($p < 0.05$) exceeds the background wtVZV exposure rate of 13% per year

Table 3. Boost and infection rates (% per year) in Oka VZV vaccinees with titers < 10 and ≥ 10U by subgroup and with varying boost definitions

Subgroup	Titer < 10 U			Titer ≥ 10 U		
	boost rate	infection rate	boost+infection rate	boost rate	infection rate	boost+infection rate
All	13.8	5.0	18.8	7.7	0.7	8.4
Year 1	13.9	4.2	18.1	8.2	0.6	8.8
Year 2	13.3	4.9	18.2	5.8	0.8	6.5
Year 3	13.2	7.1	20.3	7.3	0.4	7.7
Year 4	16.7	7.7	24.3	9.4	2.1	11.5
Age 1–4	12.9	3.4	16.3	7.3	0.7	8.1
Age 5–6	18.2	8.0	26.1	9.3	0.3	9.6
Age ≥ 7	15.9	3.3	19.2	8.4	0.2	8.7
Boost => 10 fold	9.9	5.0	14.9	4.2	0.7	4.9
Boost => 8 fold	10.3	5.0	15.3	4.8	0.7	5.5
Boost => 2 fold	24.3	5.0	29.3	13.8	0.7	14.5

When comparing subjects with high versus low anti-VZV titers (Table 3), the excess infection plus boost rate in the latter population was observed during each year of follow-up (2.0–2.8-fold, $\chi^2 > 7.5$, $p < 0.01$) and persisted even if the definition of immunologic boosting was changed to encompass antibody titer increases anywhere in the range from > 2-fold to > 10-fold (2.0–3.0-fold, $\chi^2 > 134$, $p < 0.001$). This was also true for all age groups (2.0–2.7-fold, $\chi^2 > 16.0$, $p < 0.001$), including children aged 1–4 (whose wtVZV exposure rate was lower than that of older children in previous studies [6, 9, 28]). The infection plus boost rate in children with titers < 10U also significantly exceeded the background wtVZV exposure rate of 13% per year even if the data were stratified for subject age, year post vaccination, or definition of a boost (Table 3, $z_c > 2.5$, $p < 0.01$ for all subgroups).

The excess boosts could not be attributed to increased assay variability among subjects with low titers. The standard deviation of the gpELISA assay was 0.3U. If low anti-VZV levels were responsible for increased assay variability, the frequency of a > 4-fold titer decrease should also be elevated in this population. In contrast, such decreases occurred in only 3.5% of subjects with titers < 10U but were nearly twice as frequent in subjects with higher titers (whose immunity waned at a rate of 0.75 fold/yr, see above). Although unboosted titers waned at a geometric mean rate of 0.71 fold per year, the increase in antibody titer associated with boosting was so great that 100% remained above the pre-boost levels when a third consecutive titer was available (n = 149), Moreover, as noted above, > 4-fold rises in anti-VZV titer correlated with significantly improved protection from infection. Indeed, the chickenpox infection rate among boosted individuals was 0.8% per year versus a mean of 3.4% for non-boosted study participants (Table 1, Line I, $\chi^2 = 24.5$, $p < 0.001$). Of note, subjects who initially developed a strong anti-VZV response that waned to < 1.25U over the course of the study also manifested a high subsequent infection plus boost rates (33.6%/yr) similar to that of children who responded poorly to the initial immunization.

We also subjected a database containing annual titers of vaccinees immunized with hepatitis B subunit vaccine (which does not establish latent infection) to the same analysis [21]. Unlike the situation with OkaVZV immunized subjects, there was no relationship between anti-hepatitis B antibody levels and subsequent changes in antibody titer among individuals immunized with hepatitis B vaccine (data not shown). This analysis further demonstrated that the results of this study were not attributable to an artifact of the analysis methods.

A vaccinee's anti-VZV titer should not affect his/her likelihood of exposure to wtVZV (although it may influence the likelihood of clinically observed sequelae, asymptomatic immunologic boosting, or protection with no increase in titer). Our observation that the combined infection plus boost rate in subjects with low anti-VZV titers significantly exceeds the 13%/yr exposure rate to wtVZV is consistent with reactivation of the vaccine strain. The data set provides additional clinical evidence that such reactivation can occur. For example, among children with antibody titers below 1.25 U whose titer did not increase in the subsequent year, the annual chickenpox rate was 54.2 ± 6.5%/yr (n = 59, Table 1, Line L), far in excess of any plausible wtVZV exposure rate. Although disseminated, their disease was mild (median 69 lesions, range 6-300), consistent with attenuated OkaVZV being the source of the infection. Children with persistently low titers who developed chickenpox did not differ from other study participants in terms of age or time since vaccination.

Discussion

In this study, we present data indicating that OkaVZV vaccinees with low anti-VZV titers experience significant titer increases more frequently than can be explained by exposure to wtVZV. Moreover, a subset of children with persistently low titers develop mild varicella-like clinical sequelae at a rate higher than can be explained by wtVZV exposure. These events are likely to be caused by reactivation of OkaVZV.

Our analysis suggests that the maintenance of immunity in vaccinees could be associated with the selective reactivation of OkaVZV as anti-VZV immune responses decline. Hope-Simpson hypothesized that life-long immunity to wtVZV was maintained by a combination of re-exposure to wtVZV and repeated subclinical reactivation of latent virus [11]. Subclinical reactivation of wtVZV has been detected in immunocompromised patients, and is characterized by a significant boost in anti-VZV titers [17, 18] and/or the presence of virus in blood [29]. A benefit of such boosting in vaccine recipients would be its ability to extend the duration of immunity among vaccine recipients who would otherwise become susceptible to wtVZV as their antibody titers waned (Table 1).

Alternatively, the reactivation of OkaVZV to cause illness in otherwise healthy vaccinees raises the potential for long-term clinical sequelae of reactivation. Prior to this study, the only recognized cases of OkaVZV reactivation in healthy children involved the rare detection of zoster. It is possible either that the relatively mild nature of dermatomal OkaVZV zoster prevented many cases from coming to medical attention, or that reactivating OkaVZV is more likely to cause disseminated than dermatomal zoster in individuals with low anti-VZV immunity and that some of these cases were mistaken for wtVZV infection. This is consistent with the tendency of a single local OkaVZV inoculation to give rise to mild generalized varicella-like rashes in 3.8% of VZV-naive children, a rate that is higher than the 3.4% incidence of injection-site varicella-like rashes [19].

No prospective study has been conducted to identify reactivated OkaVZV as a cause of chickenpox (or disseminated zoster) in healthy vaccinees. However, we are aware of an otherwise healthy child who developed a disseminated rash 5 months after immunization where OkaVZV was identified in the lesions by sequence analysis [4]. Thus, the vaccine strain can give rise to disseminated rashes in vaccinees. Taken together, the evidence suggests that OkaVZV can reactivate to cause a chickenpox-like illness in some children who have low anti-VZV titers.

The 13%/yr rate of wtVZV exposure cannot account for the 32.2%/yr infection plus boost rate among children with low anti-VZV titers or the 54.2%/yr annual infection rate in a subset of low-titered vaccinees. Our findings imply that OkaVZV reactivates at a rate of at approximately 19%/yr in children with titers below 1.25U and 41%/yr in subjects with persistently low titers. This is consistent with combined symptomatic and asymptomatic wtVZV reactivation rates among the severely immunocompromised, which are reported at 40–60%/yr [17, 18, 29]. Because anti-VZV immunity is weaker in vaccinees than in wtVZV infected individuals, the long-term effect of OkaVZV reactivations on rates and severity of zoster and other sequelae is unknown. For example, it is unclear how OkaVZV reactivation will impact on individuals whose anti-VZV immunity wanes or who become immunosuppressed as adults. Results from zoster studies involving immunocompromised children suggest that OkaVZV should be safer than wtVZV infection [5, 10], yet prediction based on these studies is complicated because the children received repeated vaccinations and thus generally had strong anti-VZV humoral immunity, were already ill, were not studied for asymptomatic reactivations, and were followed for a relatively short period. From a public health standpoint, our findings strongly support the continued monitoring of varicella and zoster rates in OkaVZV vaccine recipients. If there is a substantial increase in the fraction of vaccinees whose titers wane over time, periodic re-vaccination may be needed to maintain protective immunity through adulthood.

Acknowledgements

This manuscript is derived from a previously published work (reference [15]). I would like to thank Dennis Klinman for his contributions to this work. The assertions herein are the private ones of the author and are not to be construed as official or as reflecting the views of the Food and Drug Administration.

References

1. Arvin A, Koropchak C, Wittek A (1983) Immunologic evidence of re-infection with varicella-zoster virus. J Infect Dis 148: 200–205
2. Asano Y, Suga S, Yoshikawa T, Kobayashi I, Tazaki T, Shibata M, Tsuzuki K, Ito S (1994) Experience and reason: twenty-year follow-up of protective immunity of the Oka strain live varicella vaccine. Pediatrics 94: 524–526
3. Bernstein HH, Rothstein EP, Watson BM, Reisinger KS, Blatter MM, Wellman CO, Chartrand SA, Cho I, Ngai A, White CJ (1993) Clinical survey of natural varicella compared with breakthrough varicella after immunization with live attenuated Oka/Merck varicella vaccine. Pediatrics 92: 833–837
4. Brunell PA, Argaw T (2000) Chickenpox due to vaccine virus contracted from a vaccinee with zoster. Pediatrics 106: E28
5. Brunell PA, Taylor J, Geiser CF, Frierson L, Lydick E (1986) Risk of herpes zoster in children with leukemia: varicella vaccine compared with history of chickenpox. Pediatrics 77: 53–65
6. Choo PW, Donahue JG, Manson JE, Platt R (1995) The epidemiology of varicella and its complications. J Infect Dis 172: 706–712
7. Cohen JI, Brunell PA, Straus SE, Krause PR (1999) NIH conference: recent advances in varicella-zoster virus infection. Ann Intern Med 130: 922–932
8. Colton T (1974) Statistics in medicine. Little, Brown & Co., Boston
9. Finger R, Hughes J, Meade B, Pelletier A, Palmer C (1994) Age-specific incidence of chickenpox. Pub Health Rep 109: 750–755
10. Hardy I, Gershon AA, Steinberg SP, LaRussa P (1991) The incidence of zoster after immunization with live attenuated varicella vaccine. N Engl J Med 325: 1545–1550
11. Hope-Simpson RE (1965) The nature of herpes zoster: a long-term study and a new hypothesis. Proc R Soc London 58: 9–20
12. Johnson CE, Stancin T, Fattlar D, Rome LP, Kumar ML (1997) A long-term prospective study of varicella vaccine in healthy children. Pediatric 100: 761–766
13. Keller PM, Lonergan K, Neff BJ, Morton DA, Ellis RW (1986) Purification of individual varicella-zoster virus (VZV) glycoproteins gpI, gpII, and gpIII and their use in ELISA for detection of VZV glycoprotein-specific antibodies. J Virol Methods 14: 177–188
14. Krause PR, Klinman DM (1995) Efficacy, immunogenicity, safety, and use of live attenuated chickenpox vaccine. J Pediatrics 127: 518–525
15. Krause PR, Klinman DM (2000) Varicella vaccination: evidence for frequent reactivation of the vaccine strain in healthy children. Nature Med 6: 451–454
16. Kuter B, Weibel R, Guess H, Mattthews H, Morton D, Neff B, Provost P, Watson B, Starr S, Plotkin S (1991) Oka/Merck varicella vaccine in healthy children: final report of a 2-year efficacy study and 7-year follow-up studies. Vaccine 9: 643–647
17. Ljungman P, Lonnqvist B, Gahrton G, Ringden O, Sundqvist V, Wahren B (1986) Clinical and subclinical reactivation of varicella-zoster virus in immunocompromised patients. J Infect Dis 153: 840–847
18. Luby J, Ramirez-Ronda C, Rinner S, Hull A, Vergne-Marini P (1977) A longitudinal study of varicella-zoster virus infections in renal transplant recipients. J Infect Dis 135: 659–663
19. Merck & Co. (1998) Varivax [varicella virus vaccine live (Oka/Merck)] package circular In: 52nd Physician's desk reference. Medical Economics Co., Montvale, pp 1762–1765
20. Plotkin SA (1994) Vaccines for varicella-zoster virus and cytomegalovirus: Recent progress. Science 265: 1383–1385
21. Scheiermann N, Gesemann M (1993) Anti-HBs antibody kinetics – A 4 year follow-up after hepatitis B vaccination. Zbl Bakt 278: 120–126

22. Straus SE, Reinhold W, Smith HA, Ruyechan WT, Henderson DK, Blaese RM, Hay J (1984) Endonuclease analysis of viral DNA from varicella and subsequent zoster infections in the same patient. N Engl J Med 311: 1362–1364
23. Tomita H, Tanaka M, Kukimoto N, Ikeda M (1988) An ELISA study on varicella-zoster virus infection in acute peripheral facial palsy. Acta Otolaryngol (Stockh) [Suppl] 446: 10–16
24. Varis T, Vesikari T (1996) Efficacy of high-titer live attenuated varicella vaccine in healthy young children. J Infect Dis [Suppl 3] 174: S330–334
25. Wasmuth EH, Miller WJ (1990) Sensitive ELISA for antibody to varicella-zoster virus using purified VZV glycoprotein antigen. J Med Virol 32: 189–193
26. Watson BA, Starr SE (1994) Commentary: varicella vaccine for healthy children. Lancet 343: 928–929
27. Weibel RE, J NB, Kuter BJ (1984) Live attenuated varicella vaccine: efficacy trial in healthy children. N Engl J Med 310: 1409–1415
28. Wharton M (1996) The epidemiology of varicella-zoster virus infections. Infect Dis Clin North Am 10: 571–581
29. Wilson A, Sharp M, Koropchak CM, Ting SF, Arvin AM (1992) Subclinical varicella-zoster virus viremia, herpes zoster, and T lymphocyte immunity to varicella-zoster viral antigens after bone marrow transplantation. J Infect Dis 165: 119–126

Author's address: Dr. P. R. Krause, Bldg 29A Rm 1 C 16, Division of Viral Products, CBER/FDA, Bethesda, MD 20892, U.S.A.

Investigation of varicella-zoster virus variation by heteroduplex mobility assay

W. Barrett-Muir[1], **K. Hawrami**[1], **J. Clarke**[2], and **J. Breuer**[1]

[1]Department of Medical Microbiology, St Bartholomews and the Royal London Hospitals Medical Schools, Queen Mary and Westfield College, London, U.K.
[2]Department of Cell Biology and Experimental Pathology, Institute of Cancer Research, Sutton, Surrey, U.K.

Summary. Heteroduplex mobility assays of 37 regions were performed on ten UK isolates of varicella zoster virus, four from cases of zoster, and six from cases of chickenpox. The variation between isolates was found to be 0.061%, which is at least five times lower than any other member of the human herpesvirus family. Fifteen of the 37 regions tested had 29 single nucleotide polymorphisms, and over half the polymorphisms were located in four gene fragments. Of the 29 SNPs, eleven were non-synonymous and these were clustered in six genes. Isolates from a child and her mother to whom she had transmitted the virus, were identical at every locus tested. All other viruses could be distinguished by a combination of SNPs and length polymorphisms of the repeat regions R1, R2 and R5.

Introduction

Despite the fact that the VZV genome was sequenced in 1986 [2], little is known about the genetic variability of the virus. Geographic variation in the VZV genotype, as defined by restriction enzyme markers, and variable repeat regions, has been demonstrated [6]. These markers and polymorphic regions were found to distinguish the Oka vaccine strain from wild type viruses circulating in the US and the UK [7, 10], although the genetic basis for attenuation has yet to be determined. Furthermore, temporal as well as geographic variation in VZV genotype has been revealed. In a previous study we showed that the viruses circulating in the UK were similar to those found in the USA, with all viruses possessing the Pst 1 site in gene 38 which distinguished them from the Oka vaccine strain [10]. Overall 20% of UK viruses were also positive for a Bgl 1 restriction site in gene 54 [8]. However, 60% of strains from zoster presenting in immigrants whose primary infection had occurred in the Indian subcontinent Africa or the Caribbean were Bgl+ as compared with fewer than 10% of viruses from cases of zoster in patients brought up in the UK. [8]. In order to examine further the variation amongst strains of VZV circulating within the UK we have used heteroduplex mobility assays to identify additional polymorphisms in viruses from cases of zoster and varicella occurring in UK residents.

Materials and methods

Samples

Vesicle fluid from ten UK patients was analysed (Table 1). Four of the patients had a diagnosis of zoster and six varicella. None of the cases of case 10 from whom she had contracted varicella, was examined. The demographic data about each of the patients is shown in Table 1.

PCR and genotyping

DNA was extracted from 200 μl of each vesicle fluid sample using the QIAamp Blood Mini Kit (Qiagen Ltd Crawley, UK). Viral DNA from each sample was initially genotyped at four loci by methods previously established (Hawrami) (Table 1).

To construct a polymorphic map of the VZV genome based on the estimated the rate of nucleotide sequence variation of 0.043% among VZV isolates [14], 37 sets of primers were designed to amplify 500 bp regions at 3000 base pair intervals (Table 2). Using the heteroduplex mobility assay (HMA) and allowing for the fact that HMA does not identify polymorphisms within the terminal 50 bases, this corresponds to approximately 15000 bases (12% of the viral genome). This would allow the identification of a mean of 6 to 7 base differences between samples.

Optimal primer sequences were based on the published Dumas strain sequence [2], and were designed using the Oligo 4 primer design package (National Biosciences Inc). The location of primers is shown in Table 2. PCR was performed in 100 μl reaction volumes using 1U AmpliTaq Gold DNA polymerase enzyme (Perkin Elmer) with PCR buffer II (Perkin Elmer), 200 μM of each deoxynucleotide triphosphate (dNTP), each primer at a concentration of 0.2 μM, 1 μl of test or control DNA extract and an optimal MgCl concentration determined for each primer pair from 1 mM to 4 mM $MgCl_2$ using the positive control DNA extract. Thermal cycling was performed using a Geneamp 480 (Perkin Elmer) and consisted of an initial hot start activation at 95 °C for 12 min in order to activate the AmpliTaq Gold DNA

Table 1. Sources of clinical samples

Sample	Location	Year	Disease	R1	R2	R5	Bgl1
1 (117)	London	1994	Varicella	228	7	A	Pos
2 (188)	London	1995	Varicella	291	4	A	Pos
3 (260)	Edinburgh	1989	Zoster	291	6	A	Neg
4 (265)	Edinburgh	1989	Zoster	261	6	A	Neg
5 (306)	London	1995	Varicella	228	9	A	Neg
6 (308)	London	1995	Varicella	228	9	A	Neg
7 (317)	Belfast	1984	Zoster	276	6	A	Neg
8 (318)	Belfast	1984	Zoster	309	6	A	Neg
9 (338)	Edinburgh	1991	Varicella	264	5	A	Pos
10 (348)	London	1996	Varicella	228	8	A	Neg

Table 2. Regions of VZV genome amplified and analysed

Primer name	Position (Dumas)	size (bp)	Primer name	Position (Dumas)	size (bp)
VZV-1 F	427-	506	VZV-34 F	63291-	509
VZV-1 R	932-		VZV-34 R	-63780	
VZV-4 F	3270-	458	VZV-37 F	66787-	516
VZV-4 R	3727		VZV-37 R	-67277	
VZV-6 F	6066-	455	VZV-37 a F	66074-	508
VZV-6 R	6500-		VZV-37 a R	66526-	
VZV-7 F	8723-	487	VZV-40 F	73771-	512
VZV-7 R	9189-		VZV-40 R	74260	
VZV-10 F	12423-	493	VZV-42 F	77260-	500
VZV-10 R	12894-		VZV-42 R	77741	
VZV-11 F	15898-	500	VZV-44 F	80747-	513
VZV-12 R	16378-		VZV-44 R	81259	
VZV-13 F	19017-	476	VZV-47 F	84247-	513
VZV-13 R	19470-		VZV-48 R	84759	
VZV-15 F	21709-	522	VZV-50 F	87736-	515
VZV-15 R	22230-1		VZV-51 R	88226	
VZV-17 F	24402-	487	VZV-52 F	91226-	495
VZV-17 R	24865		VZV-52 R	91699-	
VZV-19 F	27209-	497	VZV-54 F	94172-	505
VZV-19 R	27233		VZV-54 R	-94654	
VZV-20 F	30027-	496	VZV-54*Bgl*I F+	95005-	497
VZV-20 R	30500		VZV-54*Bgl*I R+	95501-	
VZV-21 F	33497-	503	VZV-55 F	98505-	496
VZV-21 R	33999		VZV-56 R	98982-	
VZV22b F	40507-	492	VZV-62a F	105497-	502
VZV-22b R	40998		VZV-62a R	105982-	
VZV-24 F	43985-	511	VZV-62 F	107299-	527
VZV-25 R	44476-		VZV-62 R	107825	
VZV-28 F	47506-	509	VZV63 F-O*	110512-	515
VZV-28 R	47992-		VZV63 RI*	111026-	
VZV-29a F	51043-	507	VZV64 F	111512-	661
VZV-29a R	51526-		VZV64 R	112172	
VZV-29 F	52795-	542			
VZV-29 R	53315-				
VZV-30 F	56314-	513			
VZV-30 R	-56804				
VZV-32 F	59798-	503			
VZV-32 R	60278				

polymerase. This was followed by either 30 or 40 cycles of amplification consisting of 94 °C for 1 min, optimal annealing temperature (Table 2) for 1 min, then 72 °C for 1 min. This was followed by a final extension at 72 °C for 10 min and a holding temperature of 4 °C. Most gene regions were amplified for HMA using 30 cycles. However gene regions 14, 30, 31, 37 and 62a were amplified using 40 cycles in order to increase PCR product yield. In the case of gene regions 28, 29a, 47/48, 50/51 and 55/56, amplification for HMA was performed using 30 cycles, while a second amplification reaction was performed using 40 cycles to generate templates for DNA sequence analysis. PCR products (5 µl) were visualised by agarose gel electrophoresis alongside a 100 bp ladder (Gibco BRL).

Heteroduplex mobility assay

HMA was optimised using PCR products spanning a *Bgl*l site in gene 54 from a negative and a positive sample. Initially a method similar to that of Ganguly et al. [3] was used. Briefly, 7.5 µl of each PCR product was mixed and heated at 98 °C for 10 min, slowly cooled to 65 °C over a 30 min period and held at 60 °C for 15 min then cooled to room temperature (21 °C). The samples were then mixed with 7.5 µl of gel loading buffer (45% formamide, 30% ethylene glycol, 10 mM EDTA, 5% Ficoll and 0.05% Bromophenol blue and Xylene Cyanol). Samples were electrophoresised on a 6% polyacrylamide, 0.2% Methylenebisacrcylamide gel in TTE buffer containing 10 Ethanediol and 13.75% formamide. After electrophoresis the gel was fixed and stained by immersing sequentially in 1.5 l of the following solutions:

    1. 0.1% AgNO$_3$ for 15 min.

    2. 1.5% NaOH, 0.3% formaldehyde solution for approximately 5 min until bands were just visible.

    3. 5%–6% glacial acetic acid for 5 min.

    4. 1.5% NaOH for 20 minutes to aid removal of gel from glass plates.

Improved resolution of homoduplex and heteroduplex bands was achieved using a protocol (J. Clark, personal communication), in which the two PCR products (2 µl each) were mixed with 2 µl of a new gel loading buffer and denatured by heating at 98 °C for 5 min followed by 68 °C for 30 min and then held at 4 °C. By replacing Methylenebisacrcylamide with 0.6 g/L piperazine and spraying the gel plates with diluted Bind Silane it was easier to recover the gel intact.

Corresponding PCR fragments from each of the 11 isolates were mixed in equimolar amounts and denatured by heating at 98 °C for 5 min followed by 68 °C for 30 min and then a held at 4 °C. 1.5 µl of each mix was loaded on a 6% polyacrylamide gel and electrophoresed at 45 W for 2.5 to 3 h, or 4 W for 16 h, using 0.25 × Tris-Taurine-EDTA (TTE) Buffer in the upper reservoir and 1 × TTE Buffer in the lower reservoir. Homoduplex and heteroduplex bands were visualised using silver staining as described. Where a shift occurred (Fig. 1), PCR products were sequenced using the relevant PCR primers as sequencing primers in a ABI Prism dRhodamine Biosystems according to manufacturer's instructions. Cycle sequencing reactions were electrophoresed on an ABI 377 Analyser (Applied Biosystems Inc). Sequences generated were analysed using Sequence Navigator (Applied Biosystems Inc) and compared to that of the published VZV (strain Dumas) sequence [2] in order to identify sequence polymorphisms. Where a polymorphism occurred, the fragment was reamplified from the original material and resequenced to exclude PCR artefact.

Results

Thirty-seven distinct regions of the VZV genome were analysed by PCR and HMA, corresponding to approximately 12% of the genome, and used to construct a polymorphic map of the VZV genome. All 11 samples gave correctly

2 3 4 5   6 7 8   9 10 11 12 13 14 15 16 17 18 19 20 21 22 23 24 25 26 27 28 29

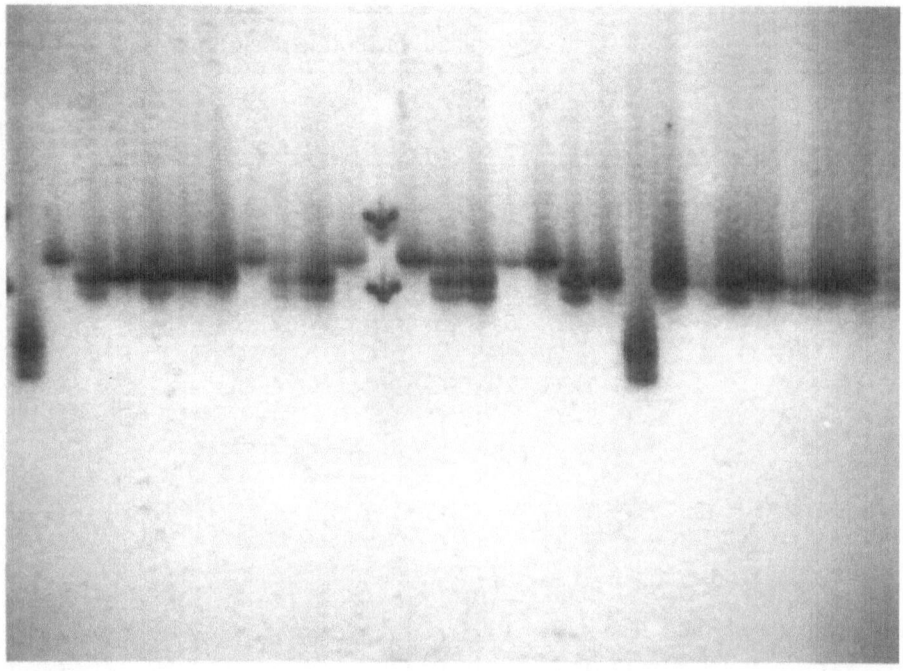

Fig. 1. Picture of heteroduplex mobility gel (HMA). *1, 13* Gibco 1 kb ladder. *2, 21* Positive heteroduplex control size 497 bp. *3, 5, 7, 8, 9, 12, 14, 17, 18, 20, 22, 23, 25, 27, 28* Homoduplex bands: 487 bp in size. *4, 6, 10, 11, 15, 16, 19, 24, 26, 29* Heteroduplex bands: 487 bp in size

sized amplification products using all 37 primer pairs. Of these 37 regions, 15 were initially identified as polymorphic by HMA. Nucleotide sequence analysis confirmed the presence of sequence variations in 14 of the HMA polymorphic regions, but not in one region (gene 28). Two regions (genes 34 and 64) gave uninterpretable HMA results Sequence analysis showed the former to contain one SNP whilst the latter contained none.

Altogether variations from the Dumas sequence were observed at twenty nine nucleotide positions within the 15 polymorphic gene regions (Table 3). The distribution of the genes in which SNPs were identified is shown in Fig. 2. Eleven SNPs resulted in non-synonymous changes in the deduced amino acid sequence, all of which occurred within six of the polymorphic gene regions (Table 4).

The rate of nucleotide substitution among the ten VZV isolates was calculated. The number of bases analysed was calculated to be the total number amplified by PCR minus the terminal 50 bases at each end of the PCR product which are not reliably analysed by HMA (15059 nucleotides), multiplied by the number of samples analysed [10]. Thus the total nucleotides analysed was 150,590. The total number of nucleotide variations from the Dumas strain sequence observed among the ten samples was 92. Thus the nucleotide variation rate was calculated to be 0.061% or one base for every 1637 nucleotides analysed.

Table 3. Nucleotide changes

Strain	Gene region and nucleotide position																											
	1	6	7	13	15	17	20	21				29a	29				34			47/48		50/51	54 *Bgl*I		55/56			
				1	2	2	2	2	3	3	3	3	5	5	5	5	5	6	6	8	8	8		9	9		9
		6	9	9	2	4	4	4	0	3	3	3	3	1	2	2	3	3	3	3	4	4	7	5	5		8
	7	7	7	7	1	0	1	0	5	5	5	6	2	6	7	7	7	3	9	9	0	1	4	6	5	6	8	2
	6	8	9	9	9	8	1	1	3	3	7	5	0	4	2	2	2	9	1	7	8	3	4	4	4	1	4	4
	6	9	0	1	8	9	3	8	3	8	4	6	6	2	5	8	0	7	7	2	3	8	1	7	6	1	1	0
Dumas	A	T	T	T	G	G	A	A	A	C	A	G	T	T	T	C	A	A	G	G	T	T	C	A	C	T	C	T
117	G	C	C	C		A			C	G	T	G		C	C	C			G	G	A			G		G	T	C
188		C	C	C		A			C	G	T	G		C	C	C								G		G	T	C
260											C											C						
265											C											C						
306					G				G	T	G		C	C	C	C	T					N			T	G	T	
308				A					G	T	G		C	C	C											G	T	
317											C											C						
318											C																	
338		C	C	C		A				G	A	C	C	C					G	G		A		G		G	T	C
348			C						G	T	G		C	C	C											G	T	

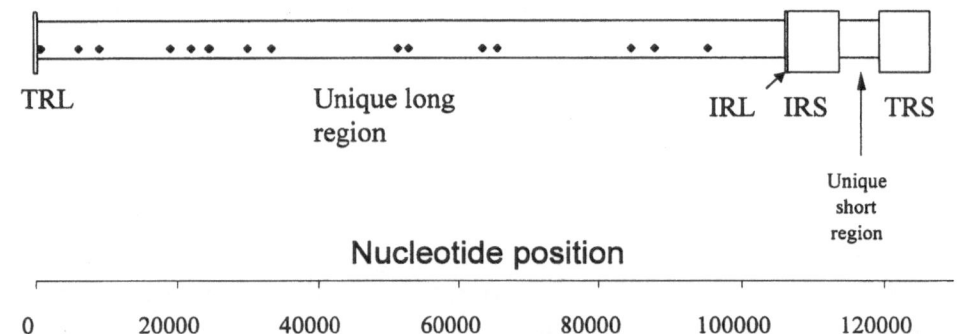

TRL   Unique long region   IRL   IRS   TRS

Unique short region

Nucleotide position

0   20000   40000   60000   80000   100000   120000

Fig. 2. Distribution of genes containing nucleotide polymorphisms identified in constructing polymorphic map. *IRS* Internal repeat short, *IRL* internal repeat long, *TRS* terminal repeat short, *TRL* terminal repeat long. ◆ Position of gene containing one or more nucleotide polymorphism

Discussion

Overall from our data the inter-strain variation of the VZV genome is 0.061%, which closely matches the estimate of Takayama et al. [14]. It has been postulated that during infection of an individual, the virus undergoes between 15 and 20 replication cycles [4]. Given that the viruses from the mother and her child were identical at all loci, this suggests that the mutation rate of the virus is extremely low i.e. less than 1 in 300,000 bases (15059 times 20). Such a low mutation rate

Table 4. Deduced amino acid variations resulting from nucleotide variations in 10 VZV samples from the UK

Strain	\n Gene region and amino acid codon position																											
	1	6	7	13	15	17		20	21				29a	29				34					47		50	54	Bgl	56
		7	1	2	1	1	1	1	9	9	9	9	1	6	7	7	7	1			4	4			2	2		
	5	4	4	9	6	2	4	2	4	6	9	6	8	8	9	7	8	0	4	5	5	9	5	8	1	4	2	8
	0	3	2	4	1	5	7	9	4	9	0	2	8	9	0	8	7	7	2	9	5	0	9	2	4	8	9	6
Dumas	I	T	Q	N	E	V	H	N	T	T	R	G	H	D	N	L	Q	Q	P	A	T	E	G	T	S	G	G	G
117	A	R							H	A	M											D					S	
188	A	R							H	A	M											D					S	
260																		A										
265																		A										
306										A	M							?										
308					I					A	M																	
317																		A										
318																												
338	A	R												E								D					S	
348			D							A	M																	

may contribute to the finding that VZV variation is considerably less than for other members of the herpesvirus family, including herpes simplex (0.32–0.81%) [12, 15] HHV8 (1.5–2%) [11], CMV (2.5%) [5] and PRV (2–3%) [9].

Despite its stability our data suggest that strains of VZV may be distinguished using single nucleotide polymorphisms and length variation of the R1, R2 and R5 repeat regions which occur within the genome. Such methods of genotyping may be useful in describing the spread and isolates. Thus the varicella isolates from the mother and her baby from whom she caught the infection were identical. By contrast, isolates, 306 and 308 (Table 1), which were isolated in the same laboratory within one month of each other from two unlinked cases of chickenpox occurring during an epidemic year, were distinguishable by three SNPs (Table 3). Similarly, viruses 260, 265 and 317, were clearly distinct at the R1 locus.

The 29 base substitutions which we found, appeared to cluster, with three of the thirty seven 500 bp gene fragments harbouring four mutations, and one containing three mutations. This apparent clustering of SNPs may have arisen by random single base mutations occurring during the evolution of the virus and this remains to be formally tested. The fact that viral strain 188 contains a subset of the sequence variants present in 117 suggests that 117 may be a descendant of 188. Similarly strains 117, 188, 306, 308 and 338 contain three or more mutations in gene 21 and may therefore be descended from strains 260, 265, 317 and 318 each of which contain only one of the base substitutions. However, as the intermediate variants between each stain and its putative descendants are not represented in this small sample, further work is underway to determine, by typing more strains, whether mutations are indeed acquired in a stepwise manner. Eleven of the nucleotide mutations observed would lead to a change in the amino acid sequence of the expressed protein. It is interesting to speculate whether such changes relate to alterations in the biological function of the virus. Recent reports have described a mutation in part of the gE glycoprotein which binds a monoclonal antibody widely used in the identification of VZV isolates [13]. The mutation abrogates binding of the monoclonal antibody and probably represents the phenomenun of antigenic drift, a property not previously associated with VZV. It is tempting to speculate that some of the amino acid changes reported here, as for example those occurring in ORF1 which codes for a membrane expressed protein which is not essential for viral growth in vitro [1], are also implicated in viral escape from immune recognition.

In conclusion, VZV is a stable virus with little variation. Viruses from single transmission events appear to be closely linked with little or no variation. Unlinked viruses are distinguishable using a combination of SNPs and length variation of repeat regions throughout the viral genome. Further work is now underway to explore aspects of VZV evolution, spread and pathogenesis using the genotyping methodology described here.

Acknowledgements

The work described in this paper was supported by grants from North Thames Health Authority Research and Development and the Special Trustees of the London and St Bartholomews Hospitals.

References

1. Cohen JI, Seidel KE (1995) Varicella-zoster virus open reading frame 1 encodes a membrane protein that is dispensable for growth of VZV in vitro. Virology 206: 835–842
2. Davison AJ, Scott JE (1986) The complete DNA sequence of varicella-zoster virus. J Gen Virol 67: 1759–1816
3. Ganguly A, Rock MJ, Prockop DJ (1993) Conformation-sensitive gel electrophoresis for rapid detection of single-base differences in double stranded PCR products and DNA fragments: Evidence for solvent-induced bends in DNA heteroduplexes. Proc Natl Acad Sci USA 90: 10325–10329
4. Grose C (1999) varicella zoster virus: less immutable than once thought. Pediatrics 103: 1027–1028
5. Haberland M, Meyer-König U, Hufert FT (1999) Variation within the glycoprotein B gene of human cytomegalovirus is due to homologous recombination. J Gen Virol 80: 1495–1500
6. Hawrami K, Harper D, Breuer J (1996) Typing of varicella zoster virus by amplification of DNA polymorphisms. J Virol Methods 57: 169–174
7. Hawrami K, Breuer J (1997) Analysis of United Kingdom wild-type strains of varicella-zoster virus: Differentiation from the Oka vaccine strain. J Med Virol 53: 60–62
8. Hawrami K, Hart IJ, Pereira F, Argent S, Bannister B, Bovill B, Carrington D, Ogilvie M, Rawstorne S, Tryhorn Y, Breuer J (1997) Molecular epidemiology of varicella-zoster virus in East London, England, between 1971 and 1995. J Clin Microbiol 35: 2807–2809

9. Ishikawa K, Tsutsui M, Taguchi K, Saitoh A, Muramatsu M (1996) sequence variation of the gC gene among pseudorabies virus strains. Vet Microbiol 49: 267–272
10. LaRussa P, Lungu O, Hardy I, Gershon A, Steinberg SP, Silverstein S (1992) Restriction fragment length polymorphism of polymerase chain reaction products from vaccine and wild-type varicella-zoster virus isolates. J Virol 66: 1016–1020
11. Poole LJ, Zong J-C, Ciufo DM, Alcendor DJ, Cannon JS, Ambinder R, Orenstein JM, Reitz MS, Haywards GS (1999) Comparison of genetic variability at multiple loci across the genomes of the major subtypes of Kaposi's sarcoma-associated herpesvirus reveals evidence for recombination and for two distinct types of open reading frame K15 alleles at the right-hand end. J Virol 73: 6646–6660
12. Sakaoka H, Kurita K, Iida Y, Takada S, Umene K, Tae Kim Y, Shang Ren C, Nahmias AJ (1994) Quantitative analysis of genomic polymorphism of herpes simplex virus type 1 strains from six countries: studies of molecular evolution and molecular epidemiology of the virus. J Gen Virol 75: 513–527
13. Santos RA, Padilla JA, Hatfield D, Grose C (1998) Antigenic variation of varicella-zoster virus Fc receptor gE: loss of a major B-cell epitope in the ectodomain. Virology 249: 21–31
14. Takayama M, Takayama N, Inoue N, Kameoka Y (1996) Application of long PCR method to identification of variations in nucleotide sequences among varicella-zoster isolates. J Clin Microbiol 34: 2869–2874
15. Umene K, Yoshida M (1993) Genomic characterization of two predominant genotypes of herpes simplex virus type 1. Arch Virol 131: 29–46

Authors' address: Dr. J. Breuer, Department of Medical Microbiology, Queen Mary and Westfield College, 37 Ashfield St, London, E1 1BB, U.K.

Varicella-zoster virus with a lost gE epitope: evidence for immunological pressure by the human antibody response

J. A. Padilla and **C. Grose**

Departments of Microbiology & Pediatrics, University of Iowa, Iowa City, Iowa, U.S.A.

Summary. The varicella-zoster virus (VZV) genome contains about 70 open reading frames (ORF). ORF 68 codes for glycoprotein gE, formerly called gp1, which is the predominant VZV glycoprotein; gE is a typical type 1 transmembrane protein with 623 amino acids. Recently, a variant virus was discovered which has a mutation in gE codon 150; this mutation converts an aspartic acid into an asparagine residue.

Introduction

Herpesviruses are considered to be ancient viruses. Based on a long series of comparative genetic analyses, investigators from the Medical Research Council Virology Unit in Glasgow have estimated that the primordial varicella-zoster virus arose some 60–70 million years before the present [14–16]. This period of evolution is of considerable interest because it includes the period when the great asteroid crashed into the Caribbean and subsequently lowered the temperature of the planet. This catastrophic event led to the demise of the dinosaurs and the ascendancy of the mammals. In fact, the first primates appeared at the same geologic period as the primordial VZV.

An important concept of the evolution of herpesviruses is that they have co-evolved with their vertebrate host. With regard to VZV, it may have arisen and evolved solely in primates [9]. Presumably, therefore, prosimians and later the great apes harbored primordial VZV. Further, about 4 million years before the present VZV followed a parallel branch of the evolutionary tree as did the earliest hominids in Africa. Finally, VZV evolved in its current form among contemporary humans. Throughout this interesting evolutionary passage, VZV has comigrated with humankind to all regions of the world. The remarkable property of latency and reactivation has allowed the virus to survive in both small and large population groups. VZV throughout the world has been considered to be a genetically similar if not virtually identical virus.

In 1998, we published a description of a VZV variant virus which had lost a B cell epitope on the glycoprotein gE ectodomain [18]. The virus has been designated VZV-MSP and the epitope has been named after the monoclonal antibody (MAb) 3B3 [10]. In this report, we provide evidence that a prominent gE immune response is produced in humans following both chickenpox and herpes zoster. Therefore, an immunological milieu exists in VZV infected humans which could lead to the selection of gE escape mutant.

This mutation lies within the B cell epitope recognized by monoclonal antibody 3B3. The amino acid change leads to a marked loss in binding of the antibody to its epitope. To determine whether immunological pressure may have contributed to the selection of this mutant virus, the humoral antibody response to gE was investigated. When the immune response to gE was determined after chickenpox, an IgG response to gE was uniformly detected. In late convalescent sera the response declined considerably; however, the anti-gE response had a dramatic anamnestic boost after herpes zoster. Finally, the anti-gE monoclonal antibody 3B3 mediated cell death in an antibody dependent cytotoxicity assay. In short, the above studies documented the presence of an anti-gE antibody response at a time when vesicle formation still was occurring in an infected child with chickenpox. Thus, these studies provided evidence for an immunological milieu which could lead to the selection of progeny virions with an absent gE epitope.

Materials and methods

Origin of cells and virus

The cell substrate for VZV propagation was the Mewo strain of human melanoma cells (HMC) [6]. Monolayer cell cultures were grown in Eagle minimal essential medium supplemented with 2 mM glutamine, 1% nonessential amino acids, penicillin (100 U/ml), streptomycin (100 μg/ml) and 10% fetal bovine serum (MEM-FBS). The VZV-32 strain of VZV has been isolated from the vesicular fluid of a young boy with chickenpox in the late 1970s [6, 8]. The VZV-MSP strain of VZV has been isolated from the vesicular fluid of a child with chickenpox in 1995 [18].

Virus infection and isotopic labeling

HMC monolayer cultures (25 cm^2 and 75 cm^2) were subcultivated at a 1:2 split ratio at 36°. Within 12 h, subconfluent cultures were seeded with one-fourth equivalent monolayer of trypsin-dispersed VZV-infected cells and incubated at 32 °C. One day later the culture medium was replaced with either MEM-FBS containing 10 μCi of tritiated fucose per ml or with methionine-deficient medium supplemented with [^{36}S]methionine (10 μCi/ml). The cultures were incubated at 32° for an additional 24 h, by which time cytopathic effect involved virtually the entire monolayer.

Polyacrylamide gel electrophoresis and fluorography

Slab gels were cast with either 8% or 10% acrylamide and cross-linked with either 2.7% methylene-bisacrylamide or an equal concentration of N,N-diallyltartardiamide. The radioactive samples were added to wells in a 4% acrylamide stacking gel and subjected to electrophoresis under constant voltage (110 V) or constant current (15 mA) in Tris-glycine buffer (pH 8.1) containing 0.1% sodium dodecyl sulfate. Upon completion of electrophoresis, the slab gels were suffused with Amplify (Amersham) prior to drying. Exposure times of the dried gels to Kodak XRP-5 film at −70° ranged from 3 to 14 days.

Immunoprecipitation

Virus-infected and uninfected HMC monolayers radiolabeled with [3H] fucose or [35S] methionine were washed and resuspended in buffer without detergents. The suspensions were disrupted by sonication and centrifuged at low speed to remove debris. Solubilization was accomplished by the addition of detergents Nonidet P-40 (NP-40) and sodium deoxycholate to final concentrations of 1% (V/V) and 1% (W/V), respectively. After a 20 min incubation, insoluble material was removed by sedimentation at 85 000 **g**. The resulting clarified extracts from VZV-infected monolayers contained $\sim 3.5 \times 10^5$ cpm of [3H]fucose and $\sim 8.8 \times 10^5$ cpm of [35S]methionine per 100 μl, while uninfected monolayers contained $\sim 6.0 \times 10^5$ and $\sim 7.7 \times 10^5$ cpm of [3H]fucose and [35S]methionine, respectively.

For analysis by immunoprecipitation, 100 μl aliquots of the solubilized antigens were incubated for 30 min at ambient temperature with 5- to 25-μl samples of the appropriate antiserum, after which the antigen-antibody mixture was incubated overnight at 4°. Preswollen protein A-sepharose CL 4B beads were added to the reaction mixture, which was held for an additional 3 h. The beads were washed 3 times with 2 ml of 0.1 M Tris-0.5 M LiCl (pH 8.8). Immune complexes were eluted at 100°C into 110 μl of sample buffer and subsequently subjected to SDS-PAGE. The molecular masses of the immunoprecipable polypeptides were estimated by comparison of their migration relative to the distance traveled by standard proteins: myosin (200,000), phosphorylase b (93,000), bovine serum albumin (68,000), ovalbumin (43,000) and cytochrome C (12,000).

Results

Presence of VZV gE in infected human tissues

The presence of VZV gE in humans was investigated by immunological methods [19]. Samples of human biopsies of vesicular lesions were probed with antibodies to VZV proteins (Fig. 1). Because of an interest in the 3B3 epitope of gE in particular, MAb 3B3 was selected as the primary antibody. In healthy humans with chickenpox, the distribution of VZV gE completely outlined the vesicle (Fig. 1). Abundant amounts of gE were located in the basal and malpighian layers of the epidermis. There was little or no immunoreactivity in the dermis underlying the infected vesicle. The presence of gE antigen was more abundant than that of other VZV antigens, such as gH (data not shown).

Humoral response of VZV gE after chickenpox

Because of the prominent production of gE antigen in human skin, the antibody response to gE was investigated in children with primary VZV infection [20]. Serum samples were collected at frequent intervals after onset of chickenpox. The antibody specificities were examined by immunoprecipitation of radiolabeled antigen prepared from VZV-infected cells; in contrast to immunoblotting, immunoprecipitation will evaluate responses to both linear and conformational gE epitopes [20]. The IgG profile during the first week was weak although antibodies to gE, gH and gB were sometimes detectable (Fig. 2). By the second week, antibodies to all glycoproteins were easily detectable. In one of the children, the IgM fraction was isolated from whole serum and shown to react also with gE, gH and gB (data not shown).

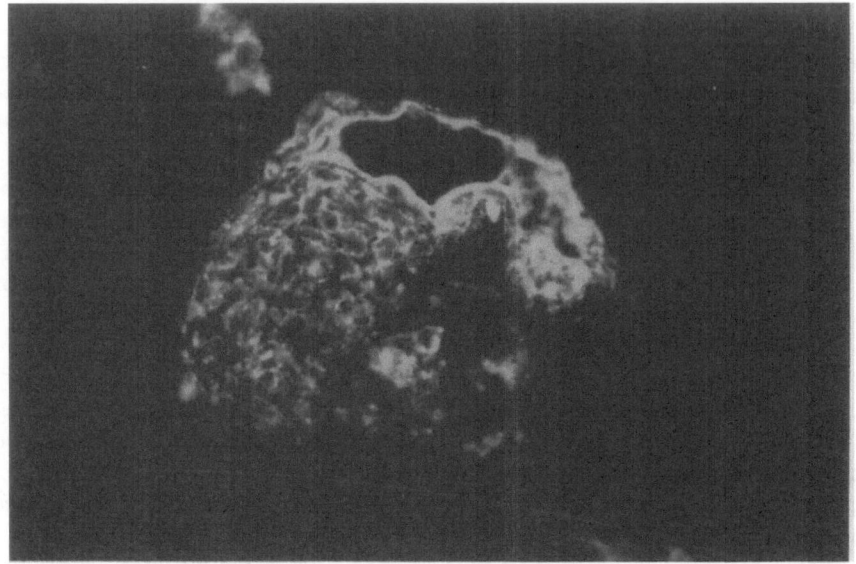
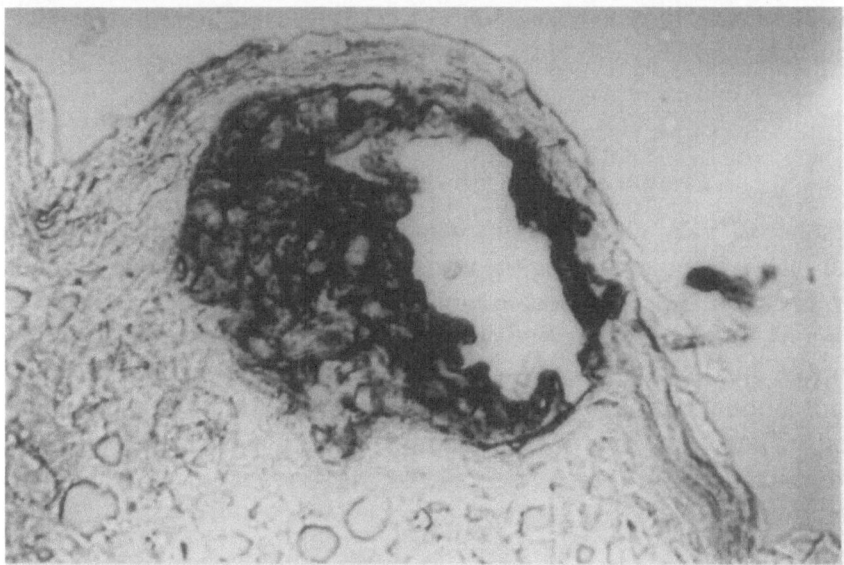

Fig. 1. Localization of VZV gE within a varicella vesicle. Biopsy samples from two varicella vesicles were probed with MAb 3B3 [10]. In the upper panel, the secondary antibody
was tagged with fluorescein. In the lower panel, the secondary antibody was tagged with
peroxidase. In both panels, gE was detected in abundant amounts in the vesicle within the
epidermis but gE reactivity stopped abruptly at the dermoepidermal junction

Humoral response to gE before and after herpes zoster

One patient with cancer developed zoster several years after primary VZV infection [7]. Serum samples collected both before and after the zoster episode were
subjected to immunoprecipitation (Fig. 3). In this patient there was a pronounced
anamnestic increase in antibody to gE following herpes zoster. There was a similar increase in anti-gH antibody. As another propitious example, serum samples

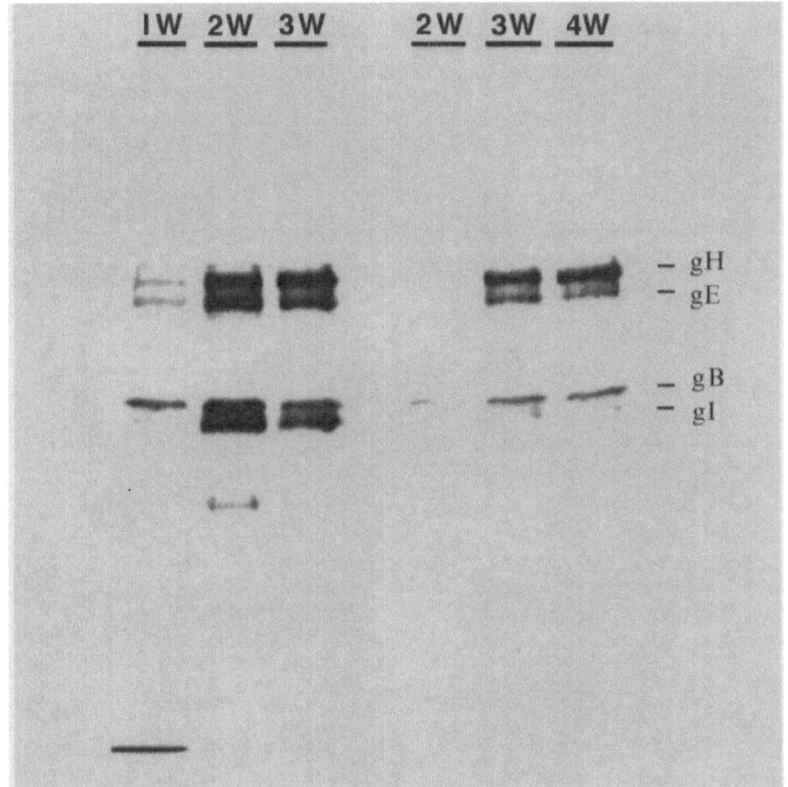

Fig. 2. Humoral response to VZV glycoproteins after primary VZV infection. Sera collected weekly (*W*) from the first patient (left) and the second patient (right) after the onset of varicella were reacted with [³H]fucose-labeled VZV antigens, and the precipitates were subjected to SDS-PAGE. As indicated by the intensity of the bands, antibodies to gE, gH and gB were detected in both patients

from a VZV immune adult were obtained immediately after his children contracted chickenpox. This individual worked in a research laboratory in the same hospital as the authors of this article; he was the parent of three small children. Prior to his intrafamilial exposure, his titers of VZV antibodies were low but detectable (Fig. 4). However, he rapidly developed an anamnestic response to the VZV glycoproteins following exposure to chickenpox in his three children. Since the parent never demonstrated any vesicular lesions, the data demonstrated either asymptomatic reinfection within the family unit or massive exposure to VZV antigens.

Summary of VZV-specific radioimmunoprecipitation profiles

The serial profiles of VZV-specific antibody responses to both glycosylated and non-glycosylated viral proteins in humans have been tabulated (Fig. 5). For the purposes of this report, the responses to gE were emphasized. The anti-gE IgG response was detectable within one week after onset of chickenpox. The responses peaked within a few weeks but remained detectable for years. Within a week

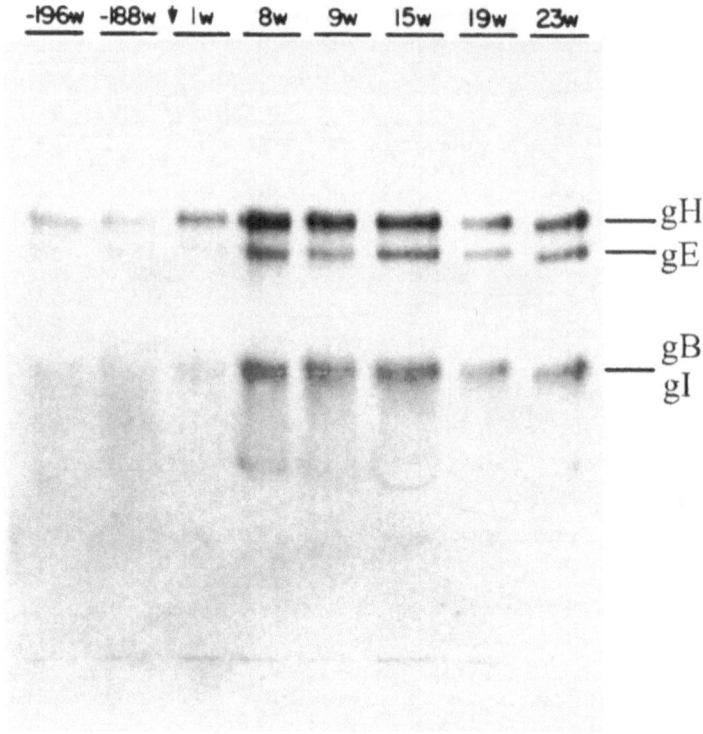

Fig. 3. VZV glycoprotein-specific antibody profiles before and after zoster. Serum samples were collected by chance 196 and 188 weeks prior to onset of zoster in one patient. After onset of zoster, samples were collected at 1, 8, 9, 15, 19 and 23 weeks

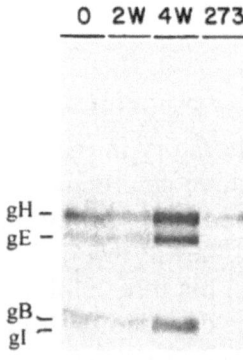

Fig. 4. Antibody pattern during subclinical reinfection with VZV. Serum samples were collected from a VZV-immune parent when his children contracted chickenpox; samples were obtained at the time chickenpox was first observed in his child (*O*) and at the indicated weeks (*W*) afterward. When the [³H]fucose-labeled glycoproteins precipitated by these sera were separated by SDS-PAGE, it was apparent that levels of antibodies to the VZV glycoproteins rose sharply after exposure to VZV

after onset of zoster, there was a marked increase in anti-gE antibody to levels higher that that seen after chickenpox. Again the anti-gE response declined over the following years but remained detectable for decades. Indeed, the antibody responses to gE, along with those to gH and gB, were among the most prominent at every phase of primary infection, latency and reactivation [20]. The profiles also are relevant to interpretation of the protective efficacy of varicella zoster immune globulin (VZIG). Since VZIG is manufactured from outdated plasma with high VZV titer, the final product will have an antibody profile similar to that

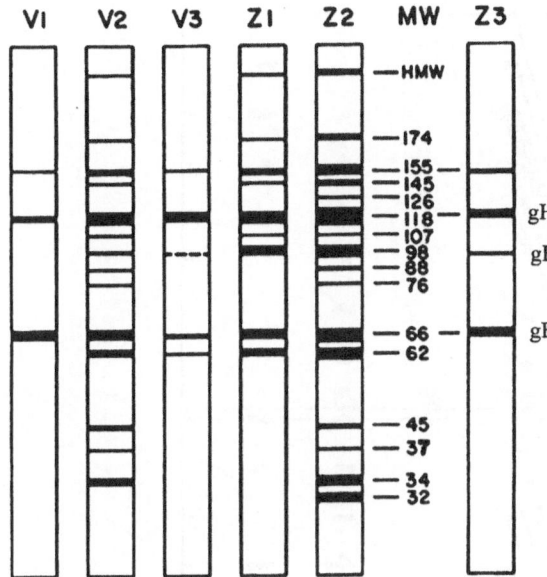

Fig. 5. Diagrammatic summary representations of the immunoprecipitable VZV polypeptides. This figure illustrates both the glycoproteins and the nonglycosylated proteins which are precipitated by human serum samples. *V1* acute-phase varicella (up to four weeks); *V2* convalescent-phase varicella (four to eight weeks); *V3* post-varicella quiescence; *Z1* acute-phase zoster (up to two weeks); *Z2* convalescent-phase zoster (two to eight weeks); and *Z3* post-zoster quiescence. The molecular masses (*MW*) are designated to the right of lane Z2. *HMW* High molecular weight protein. Adapted from [20], with permission

shown in lanes V2 and Z2; in other words, VZIG will contain large amounts of IgG antibody to the gE antigen.

Specific antibody responses to purified gE

The above studies were carried out with radiolabeled infected cell antigen preparations which contain all VZV proteins, although VZV gE is a major component. To verify and expand the above results, antibody responses to purified gE and gH were evaluated by a microtiter assay using iodinated goat anti-human IgG as a marker (Fig. 6). The wells of microtiter plates were coated with anti-gE MAb, which captured and held gE antigen for subsequent titrations of VZV gE-specific antibody in human serum samples [2]. The majority of chickenpox patients demonstrated detectable antibody to gE, with a mean acute phase titer of 200. During late convalescence, the mean titer fell to 28. After zoster, the mean titer rose dramatically, to nearly 30,000. The results regarding gH antibody were similar.

Antibody mediated cytotoxicity of VZV infected cells

The above studies conclusively demonstrated that VZV infection of humans induced a prominent polyclonal monospecific antibody response to gE. Obviously gE is a large glycoprotein with several antigenic sites harboared within its 623 amino acids. The MAb 3B3 epitope has been mapped to codons 150–161 (Table 1, peptide 3). The percentage of the polyclonal anti-gE response specifically directed toward the MAb 3B3 epitope was not determined in the prior experiments. Another assay, namely, antibody dependent cellular cytotoxicity (ADCC) was selected to investigate whether MAb 3B3 had a potential role in the immunology of VZV infection [11]. The ADCC assay was carried out with both polyclonal and monoclonal VZV-specific antibodies (Fig. 7). The results clearly showed that

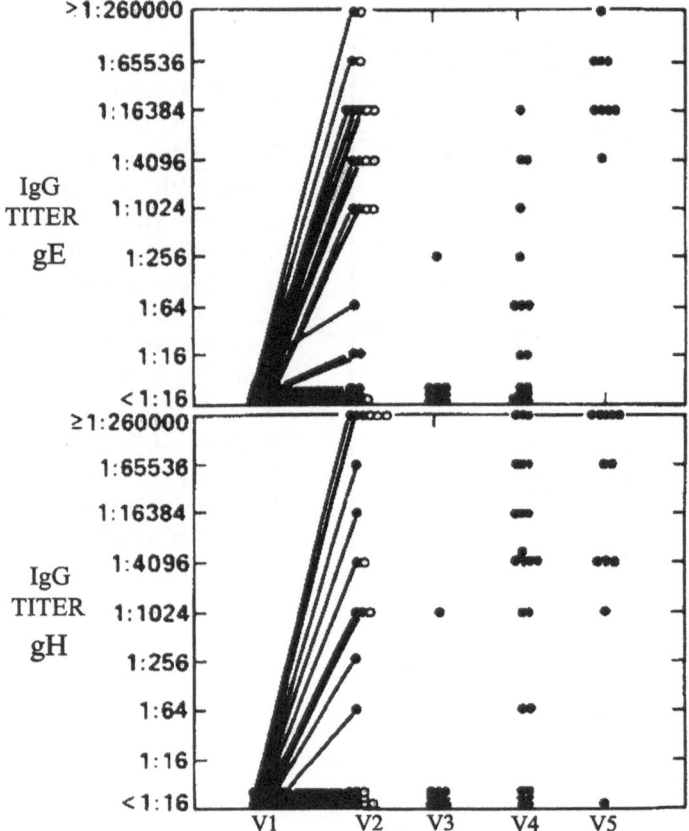

Fig. 6. Antibody responses to purified VZV gE and gH glycoproteins. The titer was expressed as the reciprocal of the serum dilution which gave a ratio of greater than 2.5 mean cpm between VZV glycoprotein and control wells of the radioimmunoassay. *V1* Acute chickenpox, *V2* convalescent chickenpox, *V3* post vaccination, *V4* known VZV seroimmune adults, *V5* convalescent herpes zoster. Adapted from [2], with permission

antibody specific for the MAb 3B3 epitope was involved in the ADCC mechanism to eliminate virus infected cells.

Discussion

Finding a VZV gE antigenic variant was completely unexpected [4, 18]. Until recently, VZV isolates around the world had been assumed to possess virtually identical immunological profiles. One ORF 10 variant virus has been reported [12], but no viruses with mutations in a surface glycoprotein have been recognized. The mutation in the VZV-MSP gE structural gene consists of a single nucleotide substitution that leads to a change in codon 150 from aspartic acid to asparagine; in turn this mutation causes a marked loss in binding of MAb 3B3 to its epitope (Table 1, peptide 3). This series of events is essentially the same phenomenon as that called antigenic drift in reference to the hemagglutinins of influenza viruses. Of note, the addition of a new asparagine residue into the gE ectodomain does not introduce an additional site for N-linked glycosylation.

Table 1. B and T cell epitopes of VZV gE

Peptide	Amino acid sequence
1. 87–101	DYDGFLENAHEHHGV
2. 103–119	NQGRGIDSGERL MOPTQ
3. 150–161	DQRQYGDVFKG
4. 181–194	PFTLRAPIQRIYGV
5. 370–381	APFDLLLEWLYV
6. 420–440	LAQRVASTVYQNCEHADNYTA
7. 562–574	KRMRVKAYRVDKS

Data obtained from [3] and [18]

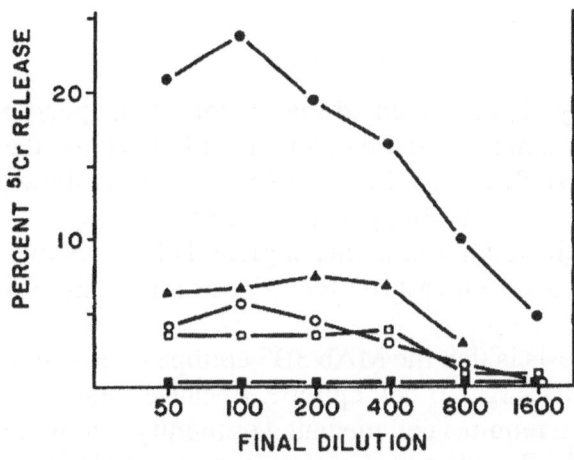

Fig. 7. Antibody dependent cellular cytotoxicity to individual VZV glyco-proteins. The cytotoxicity assays included whole VZV immune serum (●) as well as monoclonal antibodies to gE (▲), gB (O), gI (□) and gH (■). The MAb to gE was the 3B3 clone. Adapted from [11], with permission

Genetic changes within the VZV genome can be attributed to either random genetic drift or to selection. One hypothesis is that the mutation found in the VZV-MSP genome is the result of positive selection. The accepted pathogenic schema for chickenpox in a healthy child predicts that primary VZV infection consists of two viremic phases which occur over a 14-day incubation period [6]. This period includes approximately 20 replication cycles. In children with cancer the replication cycles persist for a longer period because of a diminished immune response [1]. In the case under consideration, during one of the last replication cycles in the infected child, a spontaneous mutation occurred in gE codon 150 which led to the asparagine to aspartic acid change in amino acids. The selection hypothesis predicts that the gE variant virus survived because it was less suppressed by the early anti-gE humoral response than was wild type virus. Thus the variant virus was present in the vesicular lesions and subsequently isolated by culture.

An alternative hypothesis is the mutation in gE codon 150 occurred completely by chance and was not related to survival of the variant virus. The immunology data presented in Results offer greater support to the first hypothesis because they document an early and persistent gE specific antibody response in children with

Table 2. Antibody titers to individual VZV glycoproteins measured by ELISA

Subjects	anti-gE	anti-gB	anti-gH
Healthy, before varicella	1	1	1
Healthy, after varicella	433	444	94
Cancer, before varicella	1	1	2
Cancer, after varicella	1040	422	28
Healthy, before vaccine	1	1	1
Healthy, after vaccine	33	29	16

Serum antibody titers were measured by ELISA with purified individual VZV glycoproteins as antigens. All measurements were expressed as the reciprocal genometric mean titer; in most groups, 26–34 children were tested. Data modified from [13].

chickenpox. The Gershon laboratory also has studied the glycoprotein specific antibody responses in children after natural varicella and varicella vaccine; the titers to individual glycoproteins were determined by ELISA [5, 13]. As shown in Table 2, they found a prominent anti-gE antibody response following both infection and vaccination. Of interest, the relative amounts of gE and gH antibodies may differ depending on the assay used to measure titers. For example, ELISA titers to gH are invariably low.

A third and more remote hypothesis is that the MAb 3B3 epitope contains an unrecognized T cell epitope within its larger B cell epitope. Because successful recovery from primary VZV infection requires cell mediated immunity, cytotoxic T cell epitopes certainly are biologically relevant determinants in several VZV proteins [1]. Under this scenario, the variant virus still would have been selected on the basis of immunological pressure, but the selection mechanism would involve the loss of a T cell epitope in addition to or instead of a B cell epitope. Purified gE glycoprotein is known to induce the proliferation of T lymphocytes; the gE transformation index (mean cpm in antigen stimulated and control stimulated lymphocyte cultures) is 13 in convalescent varicella and nearly 5 in people with long past (remote) infection [2]. Further, six of the T cell epitopes within gE have been mapped and their locations are shown in Table 1, peptides 1, 2 and 4–7 [3]. As is apparent, a cytotoxic T cell epitope was not mapped within the MAb 3B3 epitope (peptide 3) in an initial analysis.

In summary, gE induces a potent primary antibody response following an initial VZV infection. If by chance a mutation in gE codon 150 occurred during a late replication cycle in a child with chickenpox, the anti-gE antibody response certainly would be less able to suppress this mutant virus because it has the attributes of a monoclonal antibody escape mutant. By this means, the virus would undergo positive selection. The fact that the VZV-MSP mutant virus exhibits a different phenotype upon egress from the infected cell further suggests that the variant virus will be found to have additional biological properties, some of which may distinguish it from other wildtype strains (Fig. 8).

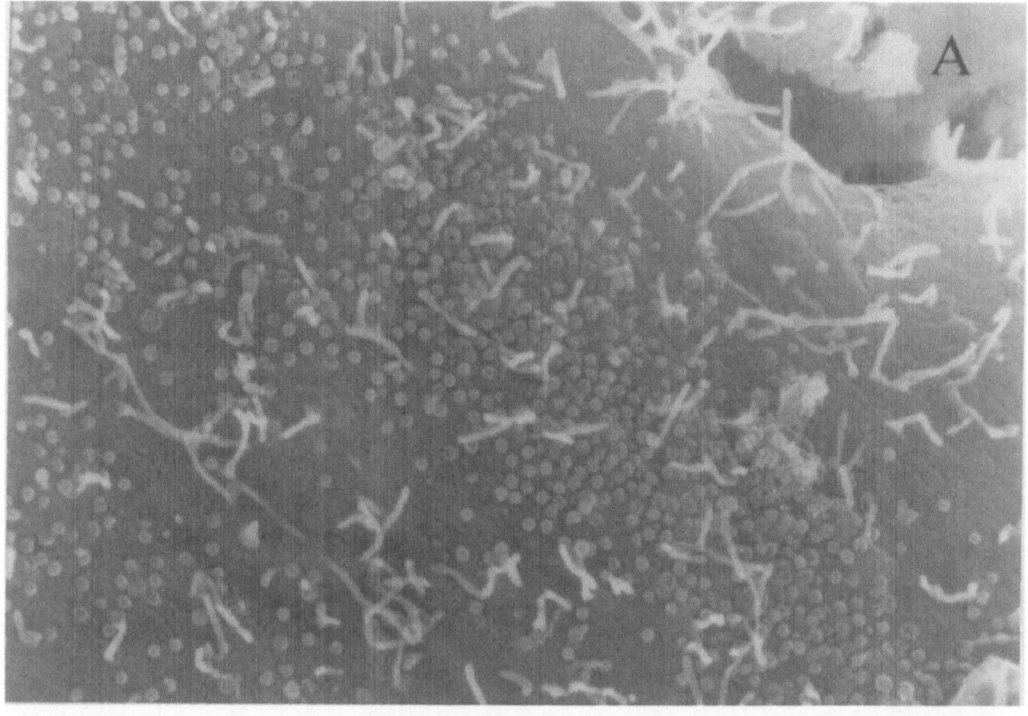

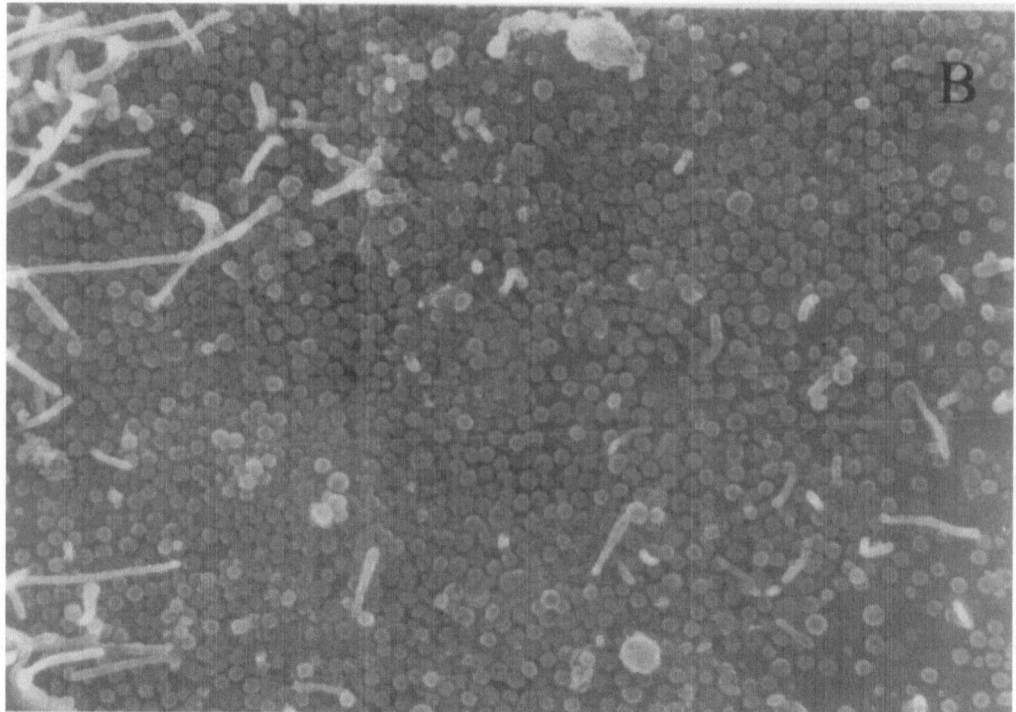

Fig. 8. Imaging of surface of VZV infected monolayers. Monolayers of melanoma cells infected with either VZV-32 or VZV-MSP strains were imaged by scanning electron microscopy. **A** shows the typical pattern of viral highways following infection with the VZV-32 strain, while **B** illustrates that the VZV-MSP strain emerges in a more uniform distribution over the cells surface. Original magnification: 10,000×

Acknowledgements

This research was supported by NIH grants AI22795 and AI36884 as well as a postdoctoral fellowship from the VZV Research Foundation of New York City.

References

1. Arvin AM (1996) Varicella-zoster virus. In: Fields BN, Knipe DN, Howley PM (eds) Virology, 3rd ed. Lippincott-Raven, Philadelphia, pp 2547–2585
2. Arvin AM, Kinney-Thomas E, Shriver K, Grose C, Koropchak CM, Scanton E, Wittek AE, Diaz PS (1986) Immunity to varicella-zoster virus glycoproteins gpI (gp 90/58) and gpIII (gp 118), and to a nonglycosylated protein, p170. J Immunol 137: 1346–1351
3. Bergen RE, Sharp M, Sanchez A, Judd AK, Arvin AM (1991) Human T-cells recognize multiple epitopes of an immediate early tegument protein (IE62) and glycoprotein I of varicella zoster virus. Viral Immunol 4: 151–166
4. Davison AJ, Scott JE (1986) The complete DNA sequence of varicella-zoster virus. J Gen Virol 67: 1759–1816
5. Gershon AA, Steinberg SP, The NIAID Varicella Vaccine Collaborative Study Group (1989) Persistence of immunity to varicella in children with leukemia immunized with live attenuated varicella vaccine. N Engl J Med 320: 892–897
6. Grose C (1981) Variation on a theme by Fenner: the pathogenesis of chickenpox. Pediatrics 68: 735–737
7. Grose C (1983) Zoster in children with cancer: radioimmune precipitation profiles of sera before and after illness. J Infect Dis 147: 47–55
8. Grose C (1990) Glycoproteins encoded by varicella-zoster virus: biosynthesis, phosphorylation, and intracellular trafficking. Annu Rev Microbiol 44: 59–80
9. Grose C (1999) Varicella-zoster virus: less immutable than once thought. Pediatrics 103: 1027–1028
10. Grose C, Edwards DP, Freidrichs WE, Weigle KA, McGuire WL (1983) Monoclonal antibodies against three major glycoproteins of varicella-zoster virus. Infect Immun 40: 381–388
11. Ito M, Ihara T, Grose C, Starr S (1985) Human leukocytes kill varicella-zoster virus infected fibroblasts in the presence of murine monoclonal antibodies of virus-specific glycoproteins. J Virol 54: 98–105
12. Kinchington PR, Turse SE (1995) Molecular basis for a geographic variation of varicella-zoster virus recognized by a peptide antibody. Neurology 45 [Suppl 8]: S13–S14
13. LaRussa PS, Gershon AA, Steinberg SP, Chartrand SA (1990) Antibodies to varicella-zoster virus glycoproteins I, II, and III in leukemic and healthy children. J Infect Dis 162: 627–633
14. McGeoch DJ, Cook S (1994) Molecular phylogeny of the Alphaherpesvirinae subfamily and a proposed evolutionary timescale. J Mol Biol 238: 9–22
15. McGeoch DJ, Cook S, Dolan A, Jamieson FE, Telford EAR (1995) Molecular phylogeny and evolutionary timescale for the family of mammalian herpesviruses. J Mol Biol 247: 443–458
16. McGeoch DJ, Davison AJ (1999) The molecular evolutionary history of the herpesviruses. In: Domingo E, Webster R, Holland J (eds) Origin and evolution of viruses. Academic Press, New York, pp 441–465
17. Padilla JA, Uno F, Yamada M, Namba H, Nii S (1997) High resolution immuno-scanning electron microscopy using a noncoating method: study of herpes simplex virus glycoproteins on the surface of virus particles and infected cells. J Electron Microsc 46: 101–110

18. Santos RA, Padilla JA, Hatfield CC, Grose C (1998) Antigenic variation of varicella zoster virus Fc receptor gE: loss of a major B cell epitope in the ectodomain. Virology 249: 21–31
19. Weigle KA, Grose C (1983) Common expression of varicella-zoster virus glycoproteins antigens in vitro and in chickenpox and zoster vesicles. J Infect Dis 148: 630–638
20. Weigle KA, Grose C (1984) Molecular dissection of the humoral response to individual varicella-zoster viral proteins during chickenpox, quiescence, reinfection and reactivation. J Infect Dis 149: 741–749

Authors' address: Dr. J. Padilla, Department of Pediatrics, University of Iowa Hospital, 200 Hawkins Drive, Iowa City, IA 52242, U.S.A.

Biologic and geographic differences between vaccine and clinical varicella-zoster virus isolates

P. S. LaRussa and **A. A. Gershon**

Columbia University, New York, New York, U.S.A.

Summary. Vaccine and wild-type strains of varicella-zoster virus differ both in their biologic characteristics and in the clinical manifestations of infection caused by each strain. The biologic differences described for the vaccine strain (temperature sensitivity and host cell preference) probably reflect the methods used to adapt the wild-type strain to the in vitro growth conditions imposed during the attenuation process in cell culture. In addition, restriction fragment polymorphisms have been described that reflect geographic strain variations between the parental virus used to develop the vaccine strain and other wild-type strains. These polymorphisms have been exploited as tools for the identification and differentiation of vaccine and wild-type strains in clinical studies. Infection with the wild-type strain results in the typical extensive rash of varicella, frequent transmission to other susceptible contacts, establishment of latency, and in some individuals, reactivation with the clinical picture of zoster. Infection with the vaccine strain results in the development of a protective immune response, minimal rash in a minority of individuals, rare transmission to other susceptible contacts, and a greatly reduced risk of zoster.

Introduction

Since Takahashi reported on the development of a live attenuated Oka varicella-zoster virus (VZV) vaccine strain in 1974 [36], much effort has been focused on identifying clinical and biologic parameters to differentiate vaccine from wild-type varicella strains. Differences in clinical parameters soon were obvious in clinical trials investigating the safety and efficacy of the Oka vaccine strain in healthy children [1–4, 36], healthy adults [13, 26, 34], and immunocompromised children [14, 35]. That is, VZV-susceptible children and adults who received the vaccine were much less likely to have a varicella-like illness in the four-week period after vaccination, as compared to susceptible children who had been exposed to wild-type varicella. Approximately 5–7% of healthy vaccinees develop a rash in the four-week period after vaccination, as compared to 95% of susceptible individuals closely exposed to the wild-type virus. When rash does occur after vaccination, it is much milder and of shorter duration (6–10 lesions for 2–3 days) than that due to the wild-type virus (mean of 300 lesions, with new lesions for up to 7 days). The vaccine strain is attenuated compared to the wild-type virus whether it is given by subcutaneous injection or by intranasal spray [5], the latter route mimicking the natural route of infection, that is, contact with the mucosa of the oropharynx. Of note, when the wild-type virus is injected by the subcutaneous route, it still results in varicella [7, 17]. Even when the vaccine virus is acquired by the natural route of contact with an immunized child with a vaccine-associated rash (see below), it remains attenuated. When rash occurs in the infected contact, it is usually mild (mean number of lesions = 38) indicating that the attenuated vaccine strain does not revert back to the wild-type phenotype even after passage in the two human hosts [37]. These observations demonstrate that the basis of the attenuated clinical response to the vaccine is not due to the route of administration. Finally, in contrast to natural infection, it has not been possible recover live vaccine virus from mononuclear cells obtained from vaccinees after immunization [33].

There also has been no definitive evidence of transmission of vaccine virus in the absence of a rash in the vaccinee. In contrast, transmission of the wild-type virus to a susceptible individual may occur after close contact with a VZV-infected person in the 24 to 48 h just before the latter develops varicella. This implies that it takes relatively little virus (either on the skin or in the oropharynx) to transmit the wild-type virus. In contrast to healthy vaccinees, children with leukemia vaccinated while on maintenance chemotherapy develop more frequent and more extensive vaccine-associated rashes. Transmission of vaccine virus from leukemic children with a vaccine-associated rash to susceptible household contacts occurred in only 23% of contacts. Of these infected contacts, 74% had a mild rash (mean = 38 lesions) and 26% had a silent seroconversion [37]. The risk of transmission was directly proportional to the number of lesions exhibited by the vaccinee. If an immunologically normal recipient of vaccine develops a rash, the risk of transmission to susceptible contacts is exceedingly low. A passive surveillance study conducted in the United States over the past five years utilizing a VZV-specific polymerase chain reaction (PCR) assay to analyze samples from potential vaccine-associated complications (see below), has documented only 3 cases of transmission of vaccine virus from a healthy vaccinee with a rash to susceptible contacts [32]. Although the design of a passive reporting system precludes an exact determination of the frequency of events studied, it does allow us to speculate about the approximate risk of transmission of vaccine virus. About 20 million doses of vaccine have been distributed in the U.S. during the period between 1995 to 1999. Because approximately 5% of healthy vaccinees develop a vaccine-associated rash during the six-week post-immunization period, there have probably been about 1 million vaccine-associated rashes during this period. These rashes were potential sources of exposure of susceptible individuals to the vaccine virus. Since we have only been able to document three cases of transmission during a time period in which there were 1 million vaccine-associated rashes, the risk of transmission from a vaccine-associated rash in a healthy vaccinee to a susceptible contact must be very low. In contrast, the risk of transmission of wild-type virus from a healthy child with varicella to susceptible household contacts is about 90% and the mean number of lesions is approximately 250–500 in those contacts that do develop varicella [28].

Finally, evidence is accumulating that the vaccine virus is less likely to cause zoster than the wild-type virus [10, 14, 23, 27, 33, 38, 39]. It is not certain whether this occurs because vaccine virus is less likely to establish latency or because it is less likely to reactivate from latency. Previous studies [10] in immunocompromised vaccinees showing that zoster is more likely in those who had either a vaccine-associated rash or breakthrough disease due to wild-type virus would seem to favor the former hypothesis.

In order to both understand the nature of the attenuation of vaccine virus and to be able to determine whether adverse events reported after vaccination were actually caused by the vaccine virus, it is important to be able to identify virologic parameters that reliably distinguish vaccine from wild-type virus strains. Compared to the wild-type strains, vaccine strain shows diminished replication in cell culture at $39\,^\circ$C vs. $34\,^\circ$C. In contrast, the vaccine strain replicates to a higher titer than the wild-type strain in guinea pig fibroblasts than in human fibroblasts. This is understandable considering that Takahashi's method of attenuation involved both 12 passages in guinea pig fibroblasts at $37\,^\circ$C and 11 passages in human embryonic lung fibroblasts at $34\,^\circ$C. The vaccine strain also expresses less of glycoprotein C (formerly gp V) than does wild-type virus [15] and is less infectious in a SCID-hu mouse human-skin explant model [25].

A number of restriction fragment polymorphisms (RFLPs) have been described that have been exploited to differentiate vaccine from wild-type virus strains. These polymorphisms, including the presence of a novel Bgl-I restriction site in gene 54 [24], the absence of a Pst-I site in gene 38 [6], and the altered mobility of the Hpa-I K fragment, appear to be due to geographic strain variation and are unrelated to the attenuation of the vaccine virus [11, 12]. These RFLPs have, however, been useful in defining the range of vaccine-associated complications [18, 20, 21]. The vaccine pattern (presence of the Bgl-I and absence of the Pst-I sites) has not been seen in wild-type varicella isolates collected in various areas of the United States and Australia, but has been seen in clinical wild-type isolates in Japan [19]. This is understandable since the Oka vaccine strain was attenuated by passage of a Japanese wild-type isolate in cell culture. Recently, polymorphisms have been described that may accurately distinguish all wild-type strains (including those of Japanese origin) from the vaccine strain [9]. Further investigations of this and other polymorphisms will determine whether they are associated with the molecular basis of attenuation of the vaccine strain.

The polymorphisms described above in genes 54 and 38 have been used to distinguish vaccine and wild-type strains in clinical isolates obtained from recipients of varicella vaccine in the United States who were thought to have vaccine-related adverse events [8, 16, 21, 22, 29, 32]. The remainder of this paper will update the findings of these studies.

Table 1. VZV specimen classification scheme

Strain I.D.	WT[a]	WT[a] (Bgl+)	Oka[b]	VZV NS[c]	VZV Negative	Inadequate
222 bp sequence amplified	Yes	Yes	Yes	±	No	No
350 bp sequence amplified	Yes	Yes	Yes	±	No	No
β-globin sequence amplified	NA[d]	NA[d]	NA[d]	Yes	Yes	No
222 bp sequence cut with Bgl-I	No	Yes	Yes	NA[d]	NA[d]	NA[d]
350 bp sequence cut with Pst-I	Yes	Yes	No	NA[d]	NA[d]	NA[d]

[a] *WT* Wild-type VZV

[b] *Oka* Oka vaccine strain VZV

[c] *NS* No strain identification was possible (VZV DNA amplified, but there were inadequate quantities of DNA for restriction analysis)

[d] *NA* Not applicable

Materials and methods

Samples were collected and VZV strain identification was performed as previously described [18]. Briefly, two sets of primer pairs were used to amplify a 222 base pair (bp) fragment in gene 54 and a 350 bp fragment in gene 38 of the VZV genome. The 222-bp product contains an asymmetrically placed Bgl-I restriction site present in the Oka vaccine strain and in approximately 15–20% of American wild-type strains. The 350 bp product contains an asymmetrically placed Pst-I restriction site present in all wild-type strains but is absent in the vaccine strain. Samples that failed to amplify with the VZV-specific primers were then subjected to a second amplification with primers specific for the human β-globin gene to determine whether there was an adequate amount of DNA available in the sample. After amplification with the VZV specific primers, restriction digestion of amplified products and separation of the digested fragments by gel electrophoresis, samples were classified according to the scheme in Table 1.

Results

In order to ensure that the polymorphisms described above were not present in wild-type strains (WT) from outside of Japan, clinical samples and stored viral isolates obtained prior to May 1994, when the Oka strain varicella vaccine became available in the United States, were studied (Table 2) [19]. All of the samples from various locations on the eastern and western coasts of the United States and Australia exhibited the wild-type genotype. The only exception was an isolate from a leukemic child in northern California who developed a vaccine-associated rash after receiving varicella vaccine as a participant in a clinical trial. The assay correctly identified this isolate as the Oka strain vaccine virus.

In a later case-control study of varicella vaccine efficacy conducted in the New Haven, Connecticut area [30, 31], 108 consecutive samples obtained between March 1997 and April 1998 from children with clinically diagnosed varicella were subjected to VZV-specific PCR analysis. All were correctly identified as the wild-type strain by the RFLP pattern that they exhibited (data not shown).

Table 2. Clinical isolates, VZV-PCR strain identification

Site of origin	N[a]	WT[b]	Oka[c]
Northern California	17	16	1
Southern California	25	25	0
Rochester, New York	25	25	0
New York, New York	26	26	0
East Australia	10	10	0
West Australia	11	11	0
Japan	9	8	1

[a]Number of samples from subjects with varicella or zoster studied by VZV-specific PCR

[b]*WT* includes all wild-type VZV strains, including Bgl[+] wild-type VZV

[c]*Oka* Oka Vaccine strain VZV

The same VZV-specific PCR assay was utilized by Sharrar and colleagues to investigate the cause of adverse events temporally related to routine vaccination of healthy children and adults in the United States [32]. Samples were obtained as part of a passive post-licensure surveillance program initiated after FDA approval of varicella vaccine in March of 1995. Although the design of the study precludes determination of the frequency of selected adverse events, it does provide useful information concerning the potential for these events to be caused by the vaccine virus.

Preliminary results from an analysis performed after the first 16 million doses of vaccine were distributed (May 1995 to April 1999) are shown in Table 3. Rashes associated with the wild-type strain occurred at a median of 8 days after vaccination while those due to the vaccine strain occurred at a median of 21

Table 3. Preliminary results, VZV vaccine passive surveillance program

Reported Adverse Event (N[a])	Oka[b]	WT[c]	VZV Negative	Inadequate sample	VZV-NS[d]
Rash within 42 days of vaccination (97)	24	38	8	19	8
Potential secondary transmission of vaccine virus (29)	3	18	5	3	0
Encephalitis/Ataxia (9)	0	2	7	0	0
Herpes Zoster (56)	22	10	4	18	2

[a]Number of samples from American recipients of varicella vaccine thought to have vaccine related adverse events

[b]*Oka* Oka vaccine strain VZV

[c]*WT* includes all wild-type VZV strains, including Bgl[+] wild-type

[d]*VZV-NS* VZV-specific PCR identified VZV DNA, but there was not enough amplified DNA to perform the restriction endonuclease analysis necessary to determine the VZV strain contained in the sample

days after vaccination. This study also showed that when the viral strain could be determined, most cases of apparent secondary transmission of vaccine virus were actually caused by the wild-type virus, probably after an unrecognized exposure to wild-type VZV. In the few cases of encephalitis and ataxia where cerebrospinal fluid (CSF) was available for analysis, 7 were negative and 2 contained the wild-type strain. Finally, analysis of samples from vaccinees with zoster has shown that both the vaccine- and wild-type strains can cause zoster. This finding indicates that the vaccine strain can occasionally establish latency and cause zoster, and that wild-type virus, probably acquired as a result of subclinical infection, can establish latency and reactivate to cause zoster despite vaccination. Other studies (see above) have shown that although the vaccine strain can cause zoster, the risk is lower in those that have received the vaccine compared to those infected with the wild-type virus.

Discussion

In summary, the clinical differences between vaccine and wild-type strains reflect the attenuated nature of the vaccine strain. The biologic differences described (temperature sensitivity and host cell preference) probably reflect the methods used to adapt the wild-type strain to the in vitro growth conditions imposed during the attenuation process in cell culture. Finally, the RFLPs described above reflect geographic strain variation rather than attenuation but have been useful tools for identification of vaccine strains in clinical studies outside Japan. It is hoped that further investigation of newly described polymorphisms will lead to a better understanding of the basis for the attenuation of the Oka vaccine strain.

Acknowledgements

This study was partly supported by NIAID # CU 51147101 and #AI 204021.

References

1. Asano Y, Nakayama H, Yazaki T, Ito S, Isomura S (1977) Protective efficacy of vaccination in children in four episodes of natural varicella and zoster in the ward. Pediatrics 59: 8–12
2. Asano Y, Nakayama H, Yazaki T, Kato R, Hirose S, Tsuzuki K, Ito S, Isomura S, Takahashi M (1977) Protection against varicella in family contacts by immediate inoculation with live varicella vaccine. Pediatrics 59: 3–7
3. Asano Y, Yazaki T, Miyata T, Nakayama H, Hirose S (1975) Application of a live attenuated varicella vaccine to hospitalized children and its protective effect on spread of varicella infection. Biken J 18: 35–40
4. Baba K, Yabuuchi H, Okuni H, Takahashi M (1978) Studies with live varicella vaccine and inactivated skin test antigen: protective effect of the vaccine and clinical application of the skin test. Pediatrics 61: 550
5. Bogger-Goren S, Baba K, Hurley P, Yabuuchi H, Takahashi M, Ogra PL (1982) Antibody response to varicella-zoster virus after natural or vaccine-induced infection. J Infect Dis 146: 260–265

6. Brunell PA, Geiser CF, Novelli V, Lipton S, Narkewicz S (1987) Varicella-like illness caused by live varicella vaccine in children with acute lymphocytic leukemia. Pediatrics 79: 922–927

7. Bruusgaard E (1932) The mutual relation between zoster and varicella. Br J Dermatol Syph 44: 1–24

8. Feder HM, Jr LaRussa P, Steinberg S, Gershon AA (1997) Clinical varicella following varicella vaccination: don't be fooled. Pediatrics 99: 897–899

9. Gomi Y, Mori T, Imagawa T, Takahashi M, Yamanishi K (1999) Virus and other clinical isolates of varicella-zoster virus. In: The International Conference on Immunity and Prevention of Herpes Zoster, Osaka, Japan, p 12

10. Hardy IB, Gershon A, Steinberg S, LaRussa P (1991) The incidence of zoster after immunization with live attenuated varicella vaccine. A study in children with leukemia. N Engl J Med 325: 1545–1550

11. Hayakawa Y, Torigoe S, Shiraki K, Yamanishi K, Takahashi M (1984) Biologic and biophysical markers of a live varicella vaccine strain (Oka): identification of clinical isolates from vaccine recipients. J Infect Dis 149: 956–963

12. Hayakawa Y, Yamamoto T, Yamanishi K, Takahashi M (1986) Analysis of varicella-zoster virus DNAs of clinical isolates by endonuclease HpaI. J Gen Virol 67: 1817–1829

13. Just M, Berger R, Luescher D (1985) Live varicella vaccine in healthy individuals. Postgrad Med J 61: 129–132

14. Kamiya H, Kato T, Isaji M, Torigoe S, Oitani K, Ito M, Ihara T, Sakurai M, Takahashi M (1984) Immunization of acute leukemic children with a live varicella vaccine (Oka strain). Biken J 27: 99–102

15. Kinchington PR, Ling P, Pensiero M, Ruyechan WT, Hay J (1990) The glycoprotein products of varicella-zoster virus gene 14 and their defective accumulation in a vaccine strain (Oka). J Virol 64: 540–548

16. Kohl S, Rapp J, LaRussa P, Gershon A, Steinberg S (1999) Natural varicella zoster virus reactivation shortly after varicella immunization in a child. Pediatr Infect Dis J 18: 1112–1113

17. Kundratitz K (1925) Experimentelle Ubertragung von Herpes Zoster auf den Mensschen und die Beziehungen von Herpes Zoster zu Varicellen. Monatssbl Kinderheilkd 29: 516–523

18. LaRussa P, Lungu O, Hardy I, Gershon A, Steinberg SP, Silverstein S (1992) Restriction fragment length polymorphism of polymerase chain reaction products from vaccine and wild-type varicella-zoster virus isolates. J Virol 66: 1016–1020

19. LaRussa P, Steinberg S, Arvin A, Dwyer D, Burgess M, Menegus M, Rekrut K, Yamanishi K, Gershon A (1998) Polymerase chain reaction and restriction fragment length polymorphism analysis of varicella-zoster virus isolates from the United States and other parts of the world. J Infect Dis 178 [Suppl 1]: S64–S66

20. LaRussa P, Steinberg S, Gershon A (1994) Diagnosis and typing of varicella-zoster virus (VZV) in clinical specimens by polymerase chain reaction (PCR). In: 34th ICAAC, Orlando, Abstract #1486

21. LaRussa P, Steinberg S, Meurice F, Gershon A (1997) Transmission of vaccine strain varicella-zoster virus from a healthy adult with vaccine-associated rash to susceptible household contacts. J Infect Dis 176: 1072–1075

22. LaRussa P, Steinberg S, Sharrar R, Galea S, Gershon A (1997) Identification and differentiation of wild (WT)- and vaccine (OKA)-strain VZV isolates by PCR from recipients of live attenuated varicella vaccine (LAVV). In: 3rd International Conference on the Varicella-Zoster Virus, Palm Beach, Florida, Abstract #P-49

23. Lawrence R, Gershon A, Holzman R, Steinberg S, NIAID Varicella Vaccine Collaborative Study Group (1988) The risk of zoster after varicella vaccination in children with leukemia. N Engl J Med 318: 543–548
24. Martin JH, Dohner DE, Wellinghoff WJ, Gelb LD (1982) Restriction endonuclease analysis of varicella-zoster vaccine virus and wild-type DNAs. J Med Virol 9: 69–76
25. Moffat JF, Zerboni L, Kinchington PR, Grose C, Kaneshima H, Arvin AM (1998) Attenuation of the vaccine Oka strain of varicella-zoster virus and role of glycoprotein C in alphaherpesvirus virulence demonstrated in the SCID-hu mouse. J Virol 72: 965–974
26. Ndumbe PM, Cradock-Watson JE, MacQueen S, Dunn H, Andre F, Davies EG, Dudgeon JA, Levinsky RJ (1985) Immunisation of nurses with a live varicella vaccine. Lancet 1: 1144–1147
27. Plotkin SA, Starr S, Connor K, Morton D (1989) Zoster in normal children after varicella vaccine. J Infect Dis 159: 1000–1001
28. Ross A, Lencher E, Reitman G (1962) Modification of chickenpox in family contacts by administration of gamma globulin. N Engl J Med 267: 369–376
29. Salzman MB, Sharrar RG, Steinberg S, LaRussa P (1997) Transmission of varicella-vaccine virus from a healthy 12-month-old child to his pregnant mother. J Pediatr 131: 151–154
30. Shapiro E, LaRussa P, Steinberg S, Gershon A (1998) Protective efficacy of varicella vaccine. In: 34th Annual Meeting of The Infectious Diseases Society of America, Denver, Abstract #78
31. Shapiro E, Vazquez M, LaRussa P, Steinberg S, Gershon A (1999) Protective efficacy of varicella vaccine. In: 24th International Herpes Virus Workshop, Boston, Abstract 13.030
32. Sharrar R, LaRussa P, Galea S, Steinberg S, Keatley R, Wells M, Stephenson W, Gershon A (2000) The postmarketing safety profile of varicella vaccine. Vaccine (in press)
33. Takahashi M (1996) The varicella vaccine. Vaccine development. Infect Dis Clin North Am 10: 469–488
34. Takahashi M, Gershon A (1994) Varicella vaccine. In: Mortimer E, Plotkin S (eds) Vaccines. WB Saunders, Philadelphia, pp 387–417
35. Takahashi M, Kamiya H, Baba K, Asano Y, Ozaki T, Horiuchi K (1985) Clinical experience with Oka live varicella vaccine in Japan. Postgrad Med 61: 61–67
36. Takahashi M, Otsuka T, Okuno Y, Asano Y, Yazaki T, Isomura S (1974) Live vaccine used to prevent the spread of varicella in children in hospital. Lancet 2: 1288–1290
37. Tsolia M, Gershon AA, Steinberg SP, Gelb L (1990) Live attenuated varicella vaccine: evidence that the virus is attenuated and the importance of skin lesions in transmission of varicella-zoster virus. National Institute of Allergy and Infectious Diseases Varicella Vaccine Collaborative Study Group. J Pediatr 116: 184–189
38. White C (1992) Varicella vaccine reflux, reply to letter. Pediatrics 89: 354
39. White CJ (1996) Clinical trials of varicella vaccine in healthy children. Infect Dis Clin North Am 10: 595–608

Authors' address: Dr. P. LaRussa, Division of Pediatric Infectious Diseases, College of Physicians & Surgeons, Columbia University, PH 4 West-462, 622 West 168th St., New York, New York 10032, U.S.A.

Comparison of DNA sequence and transactivation activity of open reading frame 62 of Oka varicella vaccine and its parental viruses

Y. Gomi[1], T. Imagawa[1], M. Takahashi[2], and K. Yamanishi[3]

[1]Kanonji Institute, The Research Foundation for Microbial Diseases of Osaka University, Kanonji, Kagawa, Japan
[2]The Research Foundation for Microbial Diseases of Osaka University, Suita, Osaka, Japan
[3]Department of Microbiology, Osaka University Medical School, Suita, Osaka, Japan

Summary. When nucleotide sequences of Oka vaccine and its parental viruses of varicella-zoster virus (VZV) were compared in 5 open reading frames (ORFs) including glycoprotein C (gC) and 4 immediate-early genes, mutations were detected only in gene 62 which is one of the immediate-early genes. Compared with its parental virus, the vaccine virus contained 15 nucleotide substitutions. With the differentiation method using the simplified restriction-enzyme fragment length polymorphism analysis by *Nae* I and *Bss* HII, which was established based on the sequence analysis data in this study, the Oka vaccine virus could be distinguished from its parental virus. Studies of the regulatory activities of the ORF62 gene product (IE62) in a transient assay indicate the IE62 of the parental virus had a stronger transactivational activity than that of the vaccine virus against immediate-early, early and late gene promoters. These data suggest that gene 62 might have an important role for attenuation of VZV. This is the first report in which many substitutions of nucleotides in gene 62 of Oka vaccine virus was found, compared with that of Oka parental virus.

Introduction

Varicella-zoster virus (VZV) causes varicella (chickenpox) and herpes zoster (shingles) in humans. Live attenuated varicella vaccine, Oka strain, was originally developed in Japan [17], and is routinely used in children in Japan and other countries, including the USA and South Korea. Clinical reactions caused by this vaccine are unusual in healthy children, but cases of chickenpox or shingles can occur, and are more common after vaccination of immunocompromised individuals. Potentially these can be caused by either wild-type or vaccine virus. Therefore, differentiation of Oka strain from other wild-type viruses is epidemiologically important. In order to distinguish Oka vaccine virus from other isolates, restriction endonuclease digestion of extracted purified viral DNA has been used [4]. At present the differentiation of clinical isolates is most often achieved using DNA amplification by polymerase chain reaction (PCR) and then characterization of amplified products by specific restriction enzyme fragment length polymorphism (RFLP) [9]. We previously reported a method for distinguishing the Oka strain from other isolates by combination analysis with the single strand conformational polymorphism of repeat region 2 and with *Pst* I-cleavage of the *Pst* I-Site-Less-region [13]. Although Oka strain can be distinguished from other isolates of VZV using methods described above, vaccine virus cannot reliably be distinguished from its parental virus by this combination method. In this paper, we found mutations in ORF 62 of Oka vaccine, suggesting that Oka vaccine virus can be differentiated not only from other VZV isolates but also from Oka parental virus by analysis of gene 62. Furthermore, we compared the transactivational activities of IE62 of vaccine virus and its parental virus.

Materials and methods

Cells and viruses

The sources of VZV strains analyzed in this study are the same as those described previously [13]. All viruses were isolated from patients with varicella or zoster in Japan who had never been vaccinated. Almost all viruses were isolated by inoculation of vesicle fluid of patients to human embryonic lung (HEL) cells and propagated by 3 to 5 passages in HEL cells. Some viruses were propagated by inoculation of materials to MRC-5 cells with 2 to 3 passages in these cells. The Oka parental virus was also propagated in MRC-5 cells. Live attenuated varicella vaccine *BIKEN* (manufactured by the Research Foundation for Microbial Diseases of Osaka University) was used as Oka vaccine virus.

Sequencing of PCR products of genes 4, 14, 61, 62 and 63

Genomic DNAs were extracted from infected cells described previously [13]. The set of primer pairs was designed by reference to the nucleotide sequence of the Dumas strain [2]. The entire gene 62 was amplified as three overlapping pieces using 3 sets of primer-pair about 40 bp in length, whose sequences are shown as follows; G62N1:5′-TTTCCCAGTCACGACGTTGTTCATAAAAACCGTTCCGC-3′ and G62R1: 5′-GGATAACAATTTCACACAGGTTCTGATCATCTACGATCCG-3′, G62N2: 5′-TTTC-CCAGTCACGACGTTGCAGGCACAACCGGTTACTCAG-3′ and G62R2: 5′-GGATAAC-AATTTCACACAGGCAAATTCGGATGATTCGGAC-3′, and G62N3: 5′-TTTCCCAGT-CACGACGTTGTTTGGTCTTACGAATCCTCGG-3′ and G62R3: 5′-GGATAACAATTTC-ACACAGGAGCGGTCTCTCCTTAAACGC-3′. Three sense-primers (G62N1, G62N2 and G62N3) had M13 forward (−38) sequence (5′-TTTCCCAGTCACGACGTTG-3′) at the end of 5′ and three antisense-primers (G62R1, G62R2 and G62R3) had M13 reverse sequence (5′-GGATAACAATTTCACACAGG-3′) at the end of 5′. Three parts of gene 62 of vaccine and parental viruses were amplified by PCR. Optimal concentration of components of the reaction mixture were determined (200 μM [each] deoxynucleoside triphosphate, 0.3 μM each primer, extremely low amount of template DNA, 2.5 U of Ex Taq [Takara shuzo Co., Kyoto, Japan] in 50 μl of Ex Taq buffer), but only the reaction using one set of primer-pair (G62N1 and G62R1) contained with 6% dimethyl sulfoxide. Thirty cycles of amplification were performed in which each cycle consisted of denaturation step at 94 °C for 1 min, annealing step at 55 °C for 1.5 min, and extension step at 72 °C for 2 min in an oil-free thermal cycler (Perkin-Elmer Co., Norwalk, USA). Then all amplified products were purified to primer-free with QIAquick PCR Purification Kit (QIAGEN GmbH, Hilden, Germany). The direct sequencing of the purified DNA specimens were performed with Thermo Sequenase Cycle Sequencing Kit (Amersham Co., Buckinghamshire, England) using M13 forward primer or reverse primer which were labeled by fluorescent IRD-40 and were analyzed using DNA sequencer Model 4000L (LI-COR Inc., Nebraska, USA) and then compiled by GENE-TYX computer programs (Software Development, Tokyo, Japan). The genes 4, 14, 61 and 63 were also amplified as described above except that specific primers were used for PCR (data not shown). The DNA sequences were determined using the same sequencer with specific primers labeled by IRD-40 and using other sequencer, Genetic Analyzer 310 (Perkin Elmer Japan-Applied Biosystems, Chiba, Japan), with unlabeled primers.

Restriction enzyme fragment length polymorphism (RFLP) analysis

The part of gene 62 was amplified using a set of primer-pairs (G62N4: 5′-GATCAAAGC-TTAGCGCAG-3′ and G62R4: 5′-CCTATAGCATGGCTCCAG-3′), 18 bp in length. Conditions of PCR were the same as those described above except for extension time to 1 min. Three μl of each amplified product were digested sequentially with 4U of *Nae* I (Takara Shuzo Co., Kyoto, Japan) at 37 °C for 1.5 h and 4U of *Bss* HII (Takara Shuzo Co., Kyoto, Japan) at 50 °C for 1.5 h. The digested products were analyzed by electrophoresis in a 4% agarose gel (NuSieve 3:1, FMC BioProducts Co., Maine, USA) with ethidium bromide staining, and visualized with short wave UV illumination of the gel.

Plasmid construction

The reporter plasmids pIE4-CAT, pgC-CAT, pDNApol-CAT, pMDBP-CAT, pIE62-CAT and pgE-CAT contained approximately 750 bases upstream of the initiation codon of the promoter region of gene 4, 14, 28, 29, 62 and 68, respectively. To create these construct DNAs, each promoter region was amplified from Oka vaccine virus by PCR with specific primers which had *Nhe* I or *Bgl* II sites at the 5′ end. Conditions for PCR were the same as those described above, except for extension time to 1 min. The amplified products were digested with *Nhe* I and *Bgl* II, and inserted upstream of chloramphenicol acetyltransferase (CAT) gene of pCAT3-Basic plasmid (Promega, Wisconsin, USA) which had been digested in advance with corresponding enzymes. The effector plasmid pPar-IE62 was a clone containing entire gene 62 from the parental virus and used for transfection. A clone containing entire gene 62 was also selected from Oka vaccine virus and used as the effector plasmid pVac-IE62, which had 11 nucleotide changes and all eight amino acid changes compared with pPar-IE62. Each plasmid contained the cognate promoter of gene 62.

Transfection

CV1 cells, which were cultured in 35-mm plastic dish at 10^5 cells/dish, were transfected by lipofection using SuperFect (QIAGEN GmbH, Hilden, Germany). Each reaction mixture for transfection consisted of 0.25 μg of the reporter plasmid, and the same amount of either effector plasmid pPar-G62 or pVac-G62. The total amount of DNA in each transfection experiment was kept constant at 2.5 μg by adding the vector DNA pUC19. All experiments were repeated at least three times independently.

CAT assay

Cells were harvested and lysed 44 h after transfection and assayed for total protein and amount of CAT. Cells were washed three times with phosphate-buffered saline and then were lysed by lysis buffer (Roche Diagnostic Co., Tokyo, Japan). The CAT concentration of individual lysate was determined by CAT ELISA based on the sandwich ELISA principle using CAT ELISA Kit (Roche Diagnostic Co., Tokyo, Japan) and was standardized to each protein concentration, determined with Bio-Rad Protein reagent (Bio Rad Laboratories, Virginia, USA).

Results and discussion

Analysis of DNA sequencing of genes 4, 14, 61, 62 and 63

We first sequenced gene 14, gC gene, of these two viruses, because the gene product of gC has been reported to play a critical role in pathogenicity [7, 12]. However, there was no difference in nucleotide sequence of the coding and flanking regions of the wild-type and vaccine strains. Because the decrease in the production of gC in Oka vaccine virus is due to a decrease of production of the major gC transcript [10], the decrease in certain transactivational activity of Oka vaccine virus as compared with its parental virus could cause the noted differences in the production of gC transcripts. It has been reported that the immediate-early gene products of ORF 4 (IE4), 61, 62 and 63 can affect the expression of immediate-early and early genes [1, 3, 6, 15, 16]. Conflicting results, however, also have been reported concerning the transactivational activity by IE 62 to late gene promoters. It was reported that IE62 can transactivate the glycoprotein E (gE) gene promoter, and that the expression of gE promoter in Vero cells is increased synergistically by IE4 and IE62 [5]. Other groups, however, showed that IE62 alone had no effect on gB and gE gene promoters [1] and a weak effect on the gI promoter [11]. When the IE4 and IE62 were expressed in human T lymphocytes, a weak effect on gC and gI gene promoters was observed [15]. Therefore, we sequenced genes 4, 61, 62 and 63 of the vaccine and its parental viruses. However, we could not find any differences in DNA sequence except gene 62. Nucleotide sequences were determined for gene 62 of Oka vaccine and its parental virus by sequencing a region from 125 bases upstream of the transcription start site (TTTTAA) to 23 bases downstream of the polyadenylation signal site (AATAAAA). When the DNA sequences of the Dumas and Oka parental viruses were compared, they were shown to differ by 9 nucleotides. In contrast 15 nucleotide changes were found in this region between Oka parental and vaccine viruses (Fig. 1). The vaccine virus with a nucleotide change at 8 positions was a mixture of two nucleotides. Every nucleotide change was a G versus A or C versus T conversion. Nucleotide changes at 8 positions in the vaccine virus caused amino acid conversions (at positions 105310, 105356, 105544, 106262, 107252, 107599, 107797 and 108838) and 4 were silent mutations (at positions 105705, 106710, 107136 and 108111). Nucleotide changes at the other three positions were determined to be in the noncoding region (at positions 105169, 109137 and 109200). The reason nucleotide replacements accumulated in gene 62 of vaccine virus is unknown, but the replacements led to genetic studies of the pathogenicity of VZV as well as the development of genetic markers for identification of vaccine virus. Because Oka virus was attenuated in guinea pig cells and in human fibroblast cells at a low temperature [17], mutant viruses,

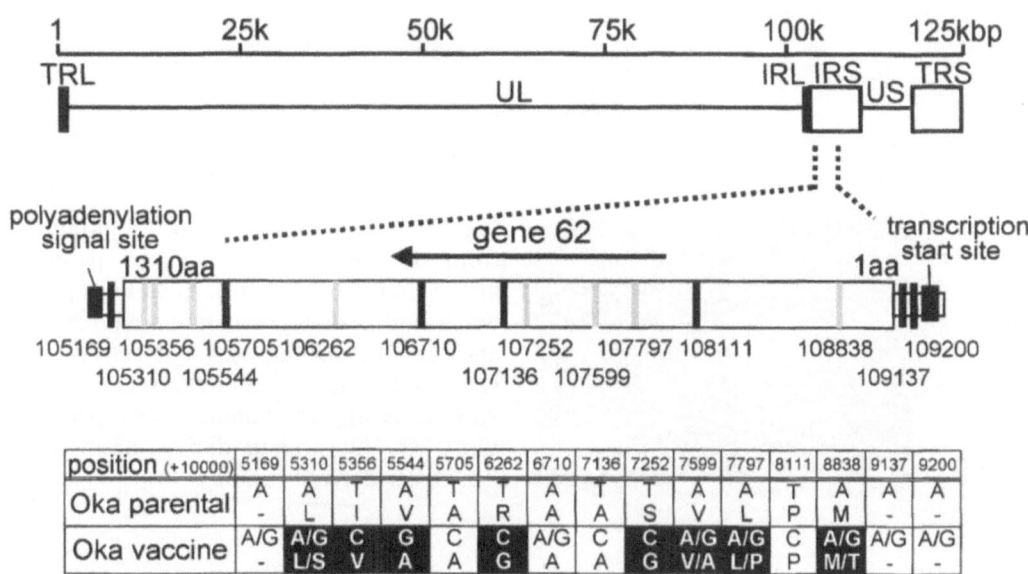

Fig. 1. Structure of gene 62 and sequence analysis of the Oka vaccine and its parental viruses. Amino acid residues are numbered from 1 to 1310 from N- to C-terminus. Vertical lines indicate the positions of nucleotide differences between the Oka vaccine and its parental viruses. Upper and lower letters in the column indicate nucleotides and amino acids, respectively.

Eight amino acid were found to be different between the viruses (in filled boxes)

which might replicate under these conditions, could have been selected during passage.

RFLP analysis

Nucleotide changes in gene 62 of Oka vaccine strain resulted in creation of new enzyme-cutting sites recognizable with restriction enzymes *Bss* HII at position 107136 (GCGCGC) and *Nae* I at position 107252 (GCCGGC). The former change was a silent mutation but the latter caused an amino acid change, i.e., Gly in vaccine virus versus Ser in the parental virus. Therefore, the effort to distinguish Oka vaccine virus from its parent by RFLP analysis was made. The PCR product, including the new cutting sites in gene 62 of Oka vaccine virus, was successfully digested with both enzymes and resulted in three fragments (400 bp, 267 bp, and 114 bp), while that of Oka parental virus was not digested by either enzyme and remained in one fragment (781 bp) (Fig. 2). RFLP study was further performed using other 54 clinical isolates derived from varicella or zoster patients. Results of PCR and RFLP analysis with these clinical isolates demonstrated that all clinical isolates tested had no cleavage-sites for both *Nae* I and *Bss* HII, resulting in one fragment of 781 bp in length as well as in Oka parental virus (Fig. 2). No significant correlation was demonstrated by the existence or absence of the *Nae* I and the *Bss* HII cleavage sites and their circulating districts, dates of isolation or source of samples (varicella or zoster). Although we have never obtained clinical isolates carrying either of the new cleavage sites for *Nae* I or *Bss* HII, further

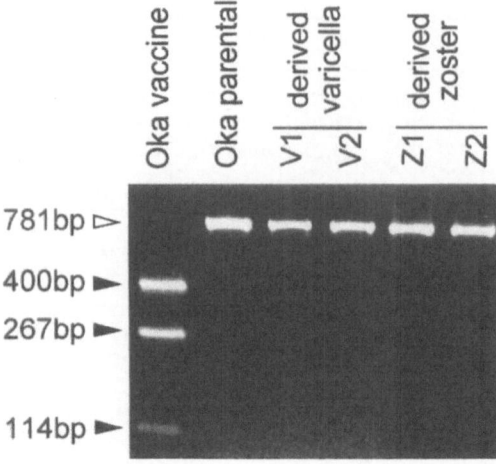

Fig. 2. RFLP analysis of PCR products from the Oka vaccine and its parental viruses and other isolates. Each PCR product was digested sequentially with *Nae* I and *Bss* HII, and analyzed by electrophoresis on a 4% agarose gel. Oka vaccine: Oka vaccine virus, Oka parental: Oka parental virus, V1–V2: varicella-derived strains, Z1–Z2: zoster-derived strains

investigation should be performed on additional clinical isolates from various sources to fully establish this method for identification of Oka vaccine virus.

Transactivational activity of IE 62 to VZV genes

To analyze the correlation between 8 amino acid mutations and transactivational activities of the IE62s, CAT assay was performed in CV1 cells transfected together with 0.25 μg of the reporter plasmid and a constant amount of the effector plasmid. All of the VZV gene promoters used displayed very low-level background activity; when no effector plasmid was transfected, little or no CAT expression was observed (Fig. 3). Two kinds of plasmids, that are promoters of immediate-early genes and cloned from vaccine virus, pCAT-IE4 and pCAT-IE62, were used for the transient CAT assay. IE62s transactivated gene 4 as well previous reports [1, 15]. The transactivational activity by parental IE 62 was 7.8-fold greater than that by vaccine IE62. However little amount of CAT could be detected from pCAT-IE62 and either effector plasmids-transfected cells (Fig. 3). Both of the IE62s were able to activate the two kinds of early gene reporters, pCAT-Pol, pCAT-MDBP, and one of late gene promoters, pCAT-gE. The activity by parental IE62 was also 7.6-fold, 5.6-fold and 1.8-fold higher than that by vaccine, respectively. Both transactivators, however, were not able to activate the other late gene reporter, pCAT-gC.

It was surprising that vaccine IE62 showed lower transcriptional activity than parental IE62 against these promoters. In fact, it was reported that the IE 62 is present in large quantity in virion tegument [8], and the yield of infectious virus generated by purified VZV DNA is greatly improved when the IE62 expressing plasmid was cotransfected [14]. Thus the mutation of vaccine IE62 may have an influence on replication of virus, resulting in an important role for attenuation of VZV.

It was demonstrated that the gC promoter remained refractory to activation by the IE4 and IE62 expressed by each cognate promoter but was weakly activated

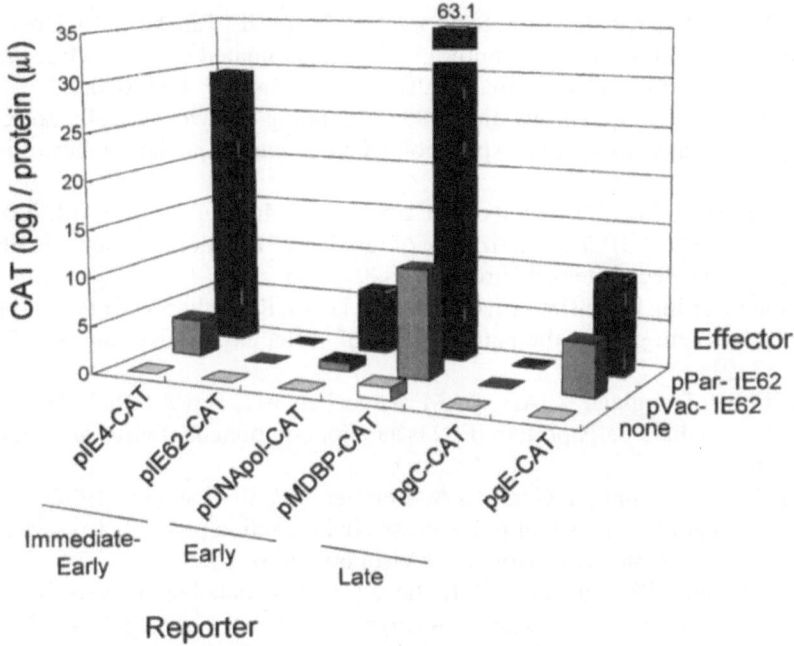

Fig. 3. Comparison of transactivational activity by the Oka vaccine and its parental IE62 on VZV gene promoters. CV1 cells were cotransfected with 0.25 μg of a reporter plasmid and same amounts of either pVac-IE62 (gray bars) or pPar-IE62 (black bars). The CAT concentrations were standardized to each protein concentration. All experiments were repeated at least three times independently

by large quantities of these proteins expressed by strong cytomegalovirus IE promoter [15]. It appears that the gC promoter was not stimulated in our experiment because the amounts of the IE62 expressed by the cognate promoter was small. We could not prove that the decrease of gC in vaccine virus was caused directly by lower transcriptional activity of the IE62. More studies are needed to clarify the correlation between pathogenicity of gC and the activity of the IE62.

Acknowledgements

We thank Drs. T. Nagai (Nagai Clinic, Takamatsu, Japan) and T. Ozaki (Showa Hospital, Aichi, Japan) for providing clinical samples of VZV.

References

1. Baudoux L, Defechereux P, Schoonbroodt S, Merville MP, Rentier B, Piette J (1995) Mutational analysis of varicella-zoster virus major immediate-early protein IE62. Nucleic Acids Res 23: 1341–1349
2. Davison AJ, Scott JE (1986) The complete DNA sequence of varicella-zoster virus. J Gen Virol 67: 1759–1816
3. Defechereux P, Melen L, Baudoux L, Merville-Louis MP, Rentier B, Piette J (1993) Characterization of the regulatory functions of varicella-zoster virus open reading frame 4 gene product. J Virol 67: 4379–4385

4. Gelb LD, Dohner DE, Gershon AA, Steinberg SP, Waner JL, Takahashi M, Dennehy PH, Brown AE (1987) Molecular epidemiology of live, attenuated varicella virus vaccine in children with leukemia and in normal adults. J Infect Dis 155: 633–640

5. Inchauspe G, Nagpal S, Ostrove JM (1989) Mapping of two varicella-zoster virus-encoded genes that activate the expression of viral early and late genes. Virol 173: 700–709

6. Jackers P, Defechereux P, Baudoux L, Lambert C, Massaer M, Merville-Louis MP, Rentier B, Piette J (1992) Characterization of regulatory functions of the varicella-zoster virus gene 63-encoded protein. J Virol 66: 3899–3903

7. Kinchington PR, Ling P, Pensiero M, Gershon A, Hay J, Ruyechan WT (1990) A possible role for glycoprotein gpV in the pathogenesis of varicella-zoster virus. Adv Exp Med Biol 278: 83–91

8. Kinchington PR, Hougland K, Arvin AM, Ruyechan WT, Hay J (1992) The varicella-zoster virus immediate-early protein IE62 is a major component of virus particles. J Virol 66: 359–366

9. LaRussa P, Lungu O, Hardy I, Gershon A, Steinberg SP, Silverstein S (1992) Restriction fragment length polymorphism of polymerase chain reaction products from vaccine and wild-type varicella-zoster virus isolates. J Virol 66: 1016–1020

10. Ling P, Kinchington PR, Ruyechan WT, Hay J (1991) A detailed analysis of transcripts mapping to varicella zoster virus gene 14 (glycoprotein V). Virology 184: 625–635

11. Ling P, Kinchington PR, Sadeghi-Zadeh M, Ruyechan WT, Hay J (1992) Transcription from varicella-zoster virus gene 67 (glycoprotein IV). J Virol 66: 3690–3698

12. Moffat JF, Zerboni L, Kinchington PR, Grose C, Kaneshima H, Arvin AM (1998) Attenuation of the vaccine Oka strain of varicella-zoster virus and role of glycoprotein C in alphaherpesvirus virulence demonstrated in the SCID-hu mouse. J Virol 72: 965–974

13. Mori C, Takahara R, Toriyama T, Nagai T, Takahashi M, Yamanishi K (1998) Identification of the Oka strain of the live attenuated varicella vaccine from other clinical isolates by molecular epidemiologic analysis. J Infect Dis 178: 35–38

14. Moriuchi M, Moriuchi H, Straus S, Cohen JI (1994) Varicella-Zoster virus (VZV) virion-associated transactivator open reading frame 62 protein enhances the infectivity of VZV DNA. Virol 200: 297–300

15. Perera LP, Mosca JD, Ruyechan WT, Hay J (1992) Regulation of varicella-zoster virus gene expression in human T lymphocytes. J Virol 66: 5298–5304

16. Schoonbroodt S, Piette J, Baudoux L, Defechereux P, Rentier B, Merville MP (1996) Enhancement of varicella-zoster virus infection in cell lines expressing ORF4- or ORF62-encoded proteins. J Med Virol 49: 264–273

17. Takahashi M, Otsuka T, Okuno Y, Asano Y, Yazaki T (1974) Live vaccine used to prevent the spread of varicella in children in hospital. Lancet 2: 1288–1290

Authors' address: Dr. K. Yamanishi, Department of Microbiology, Osaka University Medical School, 2-2 Yamada-Oka, Suita, Osaka 565-0871, Japan.

Cis and trans elements regulating expression
of the varicella zoster virus gI gene

H. He, D. Boucaud*, J. Hay, and **W. T. Ruyechan**

Department of Microbiology and Markey Center for Microbial Pathogenesis State
University of New York at Buffalo, Buffalo, New York, U.S.A

Summary. We have identified cis- and trans-acting elements involved in the VZV
IE62 protein-activated expression of the varicella zoster virus (VZV) gene which
encodes the viral gI glycoprotein. The cis-acting elements include a non-canonical
TATA box and a novel 19 base pair sequence located just upstream of the TATA
element designated the "activating upstream sequence" or AUS. The AUS is a
movable element and its presence results in IE62 activation of a chimeric promoter
consisting of the VZV gC TATA box and the gI AUS. We have also determined
that the VZV ORF 29 protein modulates the regulatory activity of the IE62 protein
at the gI promoter. In combination with the IE62 transactivator, it yields a 10 to
15-fold increase in expression over the levels seen with the IE62 protein alone in
T lymphocytes. The upmodulatory activity requires the presence of a 40 base pair
sequence, designated the 29RE, which maps between positions -220 and -180
in the gI promoter. In this paper we review these and earlier findings from our
laboratories concerning the regulation of the gI promoter.

Introduction

Varicella-zoster virus (VZV) is a member of the alphaherpesvirinae and the
causative agent of chicken pox (varicella) and shingles (zoster). Based on biochemical studies and DNA sequence analysis, the VZV genome is a linear double-
stranded DNA molecule which encodes approximately seventy proteins including
at least eight glycoproteins (gB, gC, gE, gG, gH, gI, gK, gL, and gM) [5, 11, 18,
47]. The entire complement of VZV genes is believed to be expressed during lytic
infection in three broad kinetic classes, immediate early(IE), early (E) and late
(L). Transcription of VZV genes is performed by the host cell RNA polymerase
II as is the case with all other herpesviruses [5, 47]. The transcripts are mono-

---

*Present address: Department of Molecular Genetics, M.D. Anderson Cancer Center,
University of Texas, Houston, TX 77030, U.S.A.

cistronic, polyadenylated, and predominantly, but not exclusively, unspliced [40].
Efficient expression of the VZV genome is driven by a small group of VZV gene
products including those encoded by open reading frames (ORFs) 62, 4, 61, 63
and 10 [5, 9, 10, 20, 26, 36, 37, 41–44]. The major viral transactivator is the ORF
62 protein, commonly designated IE62 [42, 44]. This protein contains a potent
N-terminal acidic transactivation domain and is capable of activating the expression of all three kinetic classes of VZV genes. ORFs 4, 61, 63, and 10 have all
been shown to have limited transactivating and/or transrepressing activities either
in the absence of, or in conjunction with, the IE62 protein.

During the latent phase of VZV infection only five ORFs are expressed [3,
6–8, 19, 32]. These include genes encoding three of the proteins known to be
involved in VZV gene expression (ORFs 4, 62, and 63), the VZV DNA single-
strand binding protein (ORF 29) and the VZV ORF 21 protein. These functions
of these gene products during latent infection is unknown, but their presence may
allow or facilitate a rapid transition to lytic infection with efficient cell to cell
spread of the virus upon reactivation.

In this report we will review our current understanding of the DNA regulatory
elements and viral and cellular proteins which regulate the expression of the gene
encoding VZV glycoprotein I (gI). Glycoprotein I (formerly designated gpIV) is
encoded by ORF 67 which is located in the unique short region of the viral genome
[11]. The protein is predicted to contain 354 amino acids and have a molecular
weight of 39, 362 kDa. However, the polypeptide chain is post-transcriptionally
modified via N-linked and O-linked glycosylation and is phosphorylated at serine
343 by a cyclin-dependent cellular kinase [18, 52]. As a result of these modifications, the fully processed form migrates with an apparent molecular weight of
55–60 kDa. By analogy with its homologue in HSV, gI is believed to be expressed
as an early/late or late gene in infected cells.

Subcellular location studies and genetic analyses have shown that gI has an
important role in the life cycle and pathogenesis of the virus. VZV gI forms a
heterodimeric complex with VZV gE which functions as an Fc receptor for non-
immune IgG [24, 30, 38]. Generation of VZV strains deleted in the gI gene results
in a major decrease in infectious virus yield, disrupted syncytium formation, a
small plaque phenotype, and a significant alteration of the intracellular localization of gE in infected cells [33]. Deletion of gI also decreases the rate of cell to
cell spread of the virus and alters the tissue tropism of VZV. Virus deleted in gI
cannot replicate in Vero cells [4] and recent work in the Arvin laboratory utilizing
a SCID-hu mouse model, has shown that gI is required for efficient growth of
the virus in human skin implants (A. Arvin, personal communication). Based on
this evidence of the importance of the gI protein in the replication, infectivity,
and pathogenesis of VZV it is clearly important to have an understanding of the
viral and cellular factors that control the expression of the gI gene. The results
reviewed here represent work performed in our laboratories over the past several
years, work that has been targeted to the identification and characterization of
proteins and viral DNA sequences involved in this process.

Materials and methods

Cells

Human foreskin fibroblasts (HFF, strain USU 521) were obtained from Monroe Vincent, Department of Pediatrics, Uniformed Services University of the Health Sciences. A3.01 cells, a CD4+ continuous human T cell line [12] were obtained from the AIDS Research and Reference Reagent Program, NIAID, NIH. HeLa cells were obtained from Dr. E. Niles and MeWo cells were obtained from Dr. C. Grose. These cell lines were grown and maintained as previously described [2, 44, 53]. PC-12 cells, a rat neuronal cell line, were obtained from the American Type Culture Collection, Rockville, MD and propagated as described by Greene et al. [17].

Plasmids

The plasmid pgI CAT, formerly designated pgpIVCAT, contained the putative gI promoter sequence (nucleotides −363 to +63 respective to the transcriptional start site of the gI gene)and along with plasmids pCMV62, pGi62, pCMV4, and p61CAT, and pgC-CAT [formerly pgpV-CAT) was constructed as described by Perera et al. [42]. The ORF 29 expression plasmid pCMV29 was constructed by inserting a 4.1 kb Sca I-Mlu I fragment containing the complete coding sequence for the ORF 29 gene into the EcoRI site of the expression plasmid pg310 containing the CMV IE promoter as described by Boucaud et al. [2]. Deletions of the gI promoter used to map the AUS sequence were generated by exonuclease III/mung bean nuclease deletion (Stratagene, La Jolla, CA) followed by insertion of the promoter fragments into the pCAT basic reporter plasmid (Promega Biotech, Madison, WI) [22, 23]. Deletions of the gI promoter used to map the 29RE element were generated using polymerase chain reaction products terminating at -220 (pgI5′3) and -180 (pgI5′4) base pairs relative to the start of transcription [1, 29].

DNA transfections and reporter gene assays

DNA transfections using HFF cells were performed by the calcium phosphate precipitation technique [15]. DNA transfections using A3.01 and HeLa cells were performed by electroporation as described by Perera et al. [42, 44] and Boucaud et al. [2]. PC-12 cells and MeWo cells were scraped, resuspended and transfected by electroporation as described by Boucaud et al. [2]. CAT Assays were performed and the results quantified as previously described [2, 42–44]. Efficiencies of transfection were spot checked by co-transfection with a plasmid expressing the E. coli β-galactosidase gene driven by the CMV IE promoter.

Results

Transcript mapping of the ORF 67 (gI) gene

Transcript mapping of the region of the VZV genome containing the ORF 67 gene has shown that three transcripts (1.7, 2.6, and 3.7 kb) are derived from ORF67 [29]. The major transcript is the 1.7 kb transcript which maps exclusively to ORF 67 and to sequences just upstream of this ORF. The 2.6 kb transcript is a minor species which is 3′ coterminal with the 1.7 kb transcript and maps continuously through ORFs 66 and 67. ORF 66 encodes one of two viral kinase activities. The 3.7 kb transcript is also a minor species and is 5′ coterminal with the 1.7 kb transcript. This transcript maps continuously through ORF 67 and 68. It is 3′ coterminal with the major transcript from ORF 68 which encodes VZV glycoprotein E. The 5′-terminus of the major 1.7 kb transcript is heterogeneous mapping 90–91 base pairs upstream from the the gI initiation codon based on primer extension analysis and is either a T or G residue.

Examination of the ORF 67 5′-untranslated region [UTR] falling between the coding regions of ORF 66 and ORF 67 reveals the presence of potential regulatory sequences which are binding sites for cellular transcription factors. These include an atypical "TATA" element ATAAAA, located -28 base pairs upstream of the transcriptional start site, and a concensus CCAAT box element located at -71 base pairs. Additional concensus CCAAT and TATA-box like sequences are located much farther upstream making them unlikely to have a primary role in transcription of the ORF 67 gene. Finally, while potential binding sites do exist for other cellular factors such as Sp1 and USF (see below), these sequences are not typical of the concensus binding sites for these factors. The situation in regard to binding sites for VZV transcriptional regulatory proteins is considerably more ambiguous. The IE62 protein has been shown to bind to sites containing the sequence 5′-ATCGT-3′ at or near the transcription start site of its own message [10, 50, 51]. Such a sequence, however, is not present in the gI promoter. Consensus binding sites for other VZV proteins known to be involved in transcription of the viral genome (ORFs 4, 61, and 63) have yet to be identified (if, in fact, they exist at all) and it is possible that other VZV proteins, as yet unidentified, could also be involved to a significant extent in viral gene expression.

Initial identification of the gI promoter and viral trans-acting factors involved in its expression

In initial efforts to determine the extent of the DNA sequences controlling expression of the ORF 67 gene, a BamH1/AccI restriction fragment encompassing sequences −363 to +63 relative to the ORF 67 transcriptional start site (Fig. 1) was cloned into a chloramphenicol acetyl transferase (CAT) reporter plasmid in both the forward and reverse orientations. Expression of CAT activity was observed in VZV-infected human foreskin fibroblast cells [HFF] from the direct but not the reverse reporter plasmids. No expression of CAT activity above basal levels was observed in uninfected cells. Primer extension analysis confirmed that the increase in CAT activity from the direct reporter plasmid (designated gI) was due to an increase in the levels of the appropriate RNA transcript and that these transcripts initiated at the same point as the authentic ORF 67 transcript [29]. Taken together, these data indicated that sequences within the upstream region selected can function as promoter elements and that these sequences require interaction with VZV-encoded protein factors and/or cellular factors induced by VZV infection. Furthermore, since only the direct orientation of the upstream sequences resulted in significant CAT expression, both the position and orientation of the cis-acting elements relative to the coding sequences are important.

In an effort to identify viral factors which are involved in activation of the ORF 67 promoter, transient transfection assays were performed using the CAT

CAT Reporter Constructs

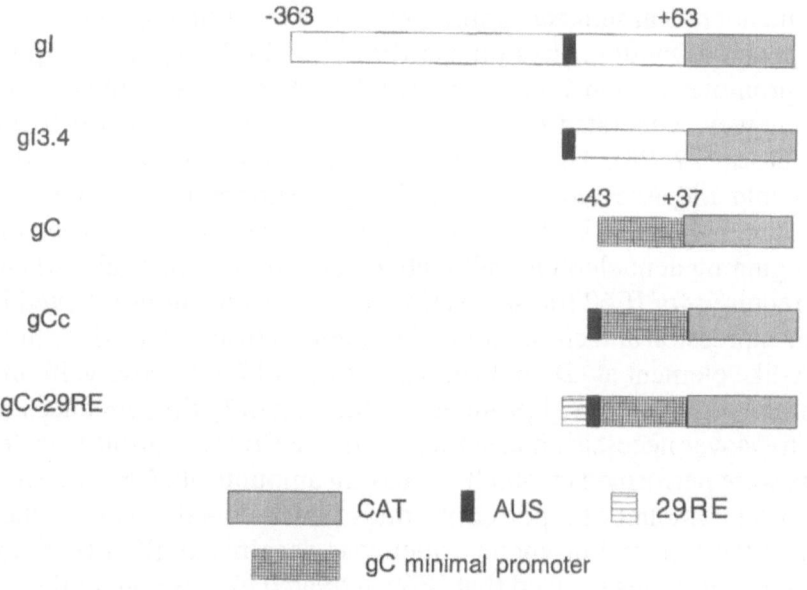

Fig. 1. Schematic diagrams of the primary reporter plasmids used in delineation of the cis-acting regulatory elements present in the gI promoter

reporter plasmid and plasmids expressing the VZV IE62 and ORF 4 proteins under the control of either their cognate promoters (pGi26 and pGORF4, respectively) or the much more powerful immediate early cytomegalovirus (CMV) promoter (pCMV62 and pCMV4). These studies showed that the IE62 protein stimulated expression from the ORF 67 promoter but to a significantly lesser extent than that observed in VZV infected cells. In contrast, the ORF 4 protein did not exhibit any ability to transactivate the ORF 67 promoter. Data obtained upon cotransfection of both effector plasmids indicated that the presence of ORF 4 increased the transactivation observed with the IE62 protein, suggesting a modulatory role for the ORF 4 protein. Significantly more activity was observed using the CMV promoter-containing ORF 62 effector plasmids as compared to that observed using the endogenous, but much weaker, VZV promoters indicating that the induction of CAT activity was dependent upon the amount of IE62 present [29]. Similar results were obtained using a continuous CD4$^+$ T cell line and Vero cells [42, 43] illustrating that the IE 62-mediated activation of the gI promoter is not cell-type dependent. This study also utilized effector plasmids expressing the ORF 61 gene product. Little or no effect on expression was observed with ORF 61 either in the presence or absence of the IE 62 and ORF 4 expressing plasmids although a later study showed a two-fold upregulation induced by ORF 61 [37]. Thus, all of the data obtained indicate that the viral protein primarily responsible for activation of the gI promoter is the VZV IE 62 protein.

Mapping of the minimal IE62 responsive gI promoter elements

As stated above, examination of the sequences upstream of the gI transcriptional start site did not reveal sequences similar to those identified by other investigators as being potential binding sites for the IE62 protein. In order to map the region of the gI promoter responsible for the IE62 mediated transactivation a series of 5′ deletions were generated using a combination of exonuclease III and mung bean nucleases [22, 23]. These truncations were cloned into the pCAT basic reporter plasmid and assessed as to their ability to support CAT expression in cells transfected with the pCMV62 expression plasmid. The results showed that the sequence beginning at nucleotide -54 relative to the transcriptional start site was all that was required for IE62 transactivation. Further truncations resulted in a sharp dropoff in expression and elimination of sequences from -35 to -13 which include the TATA-like element at -28 yielded basal levels of CAT activity. In order to determine if the region of the gI promoter containing only the sequences beginning at position -54 was necessary and sufficient for ORF 62 activation, transfection experiments were performed in which increasing amounts of pCMV62 were titrated with a constant amount of reporter plasmids which contained either the intact gI promoter or the minimal promoter which was desginated gI3.4 (Fig. 1). Results from these experiments showed that IE62 activated expression, in the absence of any other virus-encoded proteins was equivalent for both promoter constructs (He et al., manuscript in preparation). Additional deletion mapping from the 3′-end further delineated the ORF 62 responsive region to lie between positions -54 to -35. This nineteen base pair sequence, 5′-CAGAGTCACGCCCCATTAT-3′, was designated the activating upstream sequence (AUS).

One of the hallmarks of a cis-acting regulatory sequence is that it is a moveable element and can act in a heterologous setting. In order to test this in the case of the AUS sequence, it was cloned into a reporter plasmid upstream of a minimal promoter element for the VZV gC gene, pgC (Fig. 1). This sequence contains an atypical TATA element and is minimally responsive to transactivation by IE62. Transient transfection experiments with these plasmids showed that the presence of the AUS sequence correlated with ten to twenty fold-higher levels of CAT expression from the chimeric AUS-gC promoter element, gCc, as compared to those seen with the minimal gC promoter alone at a given concentration of the IE 62 expressing plasmid (H. He et al., unpublished data). Thus the AUS sequence is a movable cis-acting element which is required for efficient transactivation by the VZV IE62 protein.

Extracts derived from cells transfected with pCMV62 and a purified recombinant TrpE fusion protein containing the DNA-binding domain of IE62 [51] were both used in electrophoretic mobility gel shift assays in order to determine if the IE62 protein bound preferentially to an oligonucleotide containing the AUS sequence. No preferential binding was observed in either case as compared to controls which included mock transfected cells and an oligonucleotide of similar size and base composition. Thus the mechanism of IE62 transactivation does not appear to involve direct binding of the protein to the AUS sequence. Examination

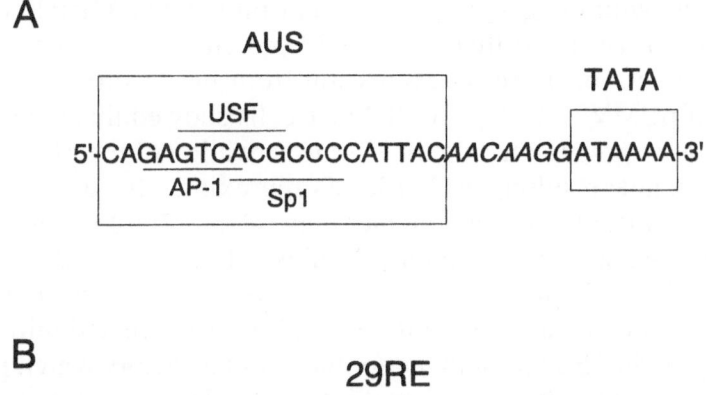

Fig. 2. Sequences of the AUS/TATA (**A**) region and the 29RE (**B**). The AUS and TATA elements are boxed. The three overlapping binding sites for cellular transcription factors within the AUS are indicated by horizontal lines

of the AUS sequence did, however, reveal the presence of binding sites for three cellular transcription factors: AP-1, USF, and Sp1 (Fig. 2A), suggesting the possibility that transactivation by IE62 involves recruitment of, or interaction with, one or more of these factors.

Finally, the ability of the AUS sequence to support transactivation by IE62 in the absence of a TATA element was assayed. The AUS sequence was inserted into the multiple cloning site of the pCAT-basic reporter plasmid. This plasmid lacks a TATA element 5′ to the CAT gene. Only a very modest increase in CAT activity above basal levels was observed upon cotransfection of this plasmid with increasing amounts of pCMV62 (less than two-fold). Thus both the AUS sequence and a TATA element were required for significant transactivation by the IE62 protein.

Modulation of IE62 transactivation by the VZV ORF29 protein

The ORF 29 protein is one of the viral genes expressed during both lytic and latent VZV infection. The ORF29 protein is the homologue of the HSV major DNA-binding protein ICP8 with which it shares 50% identity at the amino acid level. The ORF 29 protein binds preferentially to single-stranded DNA and is assumed to be indispensable for viral growth based on its predicted role in viral DNA synthesis during lytic infection [25, 46]. Its role in latency is unknown as is the case with the other gene products expressed during this phase of viral infection. However, ORFs 62, 63 and 4 have all been shown to be involved in transcription of the viral genome, and the HSV ICP8 protein has been identified as a factor influencing HSV gene expression. We therefore hypothesized that the VZV ORF 29 protein plays a role in VZV gene expression, and used the gI promoter as a model for this activity.

In experiments with the gI-CAT reporter plasmid and an ORF29 expressing plasmid (pCMV29) we found that the ORF29 protein does not appear to have any ability to increase or decrease expression from the gI promoter. However, cotransfection of pCMV29 with pCMV62 in T cells showed that the presence of the ORF 29 protein upmodulated the IE62 transactivation of the gI promoter in a dose dependent manner resulting in CAT levels approximately 16-fold higher than those observed with IE62 alone and greater than 60 to 100 fold over basal levels observed in the absence of the effector plasmids [2]. Since the ORF 29 protein binds preferentially to single-stranded DNA it is possible that it binds to the gI promoter non-specifically, destabilizing the duplex structure and allowing more efficient binding of the IE62 protein and/or the cellular factors which potentially bind to the AUS sequence. In order to test this hypothesis transfection experiments were performed in which the pCMV29 expression plasmid was substituted with the pCMVICP8 plasmid which expresses the HSV major DNA-binding protein. No upmodulation of IE62 transactivation of the gI promoter was observed in these experiments [1], (Boucaud et al. unpubl. data) indicating that the effect is specific to the VZV ORF 29 protein and not simply to the activity of a single-strand DNA binding protein, even one as closely related as ICP8.

The ORF 29 protein upmodulation was independent of the reporter gene used since expression from a luciferase reporter plasmid which contained the full gI promoter exhibited a similar increase, within experimental error, in the level of IE62 activation in the presence of the ORF 29 protein as compared to that observed with the gI-CAT reporter plasmid. In contrast to this result, the ORF29 upmodulation of IE62 activation proved to be both cell type and promoter dependent. ORF29-dependent upmodulation was also observed in human foreskin fibroblasts, MeWo, and HeLa cells. The observed effect, however, was quite modest compared to that in T cells, being no more than a two-fold increase in reporter gene activity. Data obtained from transfection of the rat neuronal PC-12 cell line showed that the presence of the ORF29 protein slightly reduced rather than increased IE62 transactivation. Thus, significant upmodulation of IE62 mediated activation of the gI promoter appears to be limited to T cells. Finally, transfection experiments using CAT and luciferase reporter plasmids containing either the SV40 IE promoter or the VZV ORF61 resulted either in no ORF 29 upmodulatory effect or a slight but reproducible decrease in IE62 activation, respectively [1, 2]. Taken together, these data suggest that the ORF 29 upmodulatory effect observed in T cells is not due simply to a direct interaction between the IE62 and ORF 29 proteins. If this were the case, the data obtained in all cell lines with the tested promoters would be similar since they are all efficiently transactivated by the IE62 protein. This is clearly not the case and additional mechanisms involving cis-acting elements and cellular factors must be considered.

Transfection experiments using the gI3.4 reporter plasmid and a reporter plasmid containing upstream sequences beginning at position -87 relative to the gI transcriptional start site did not show the ORF 29 protein upmodulation. These results suggested that addtional upstream sequences were required. In order to map the location of these sequences two additional truncations of the gI promoter were generated at positions -220 (pgI5'-3) and -180 (pgI5'-4). Transfection assays using these reporter plasmids showed that the ORF 29 protein up modulation was observed with pgI5'-3 but not pgI5'-4. The forty base pair element comprising the difference between these two promoter constructs was designated the 29 responsive element or 29RE (Fig. 2B). To test if the 29RE, like the AUS, is a mobile cis-acting element, the 29RE was placed upstream of the AUS sequence in the gCc plasmid generating the gCc29RE CAT reporter. Transfection assays comparing the ability of the ORF 29 protein to upmodulate IE62 activation of CAT expression from the two reporter plasmids showed that upmodulation was observed only with the gCc29RE promoter confirming that this 40 base pair sequence is a movable cis-acting element. The maximum upmodulation observed was approximately six-fold. Because the distance between the AUS/TATA region in this chimeric promoter is different from that in the true gI promoter, the difference in upmodulation from that observed with the gI promoter could reflect a positional as well as a sequence requirement.

Discussion

The work summarized here shows that regulation of the expression of the VZV gI gene as we currently understand it is complex and dependent on several cis and trans-acting elements. The cis-acting elements identified thus far include a TATA-box, the AUS sequence and the 29RE sequence (Fig. 3). Somewhat surprisingly, the CCAAT box located at position -73 does not appear to be a significant factor under the assay conditions and in the cell lines used. The TATA-box does not fit the typical TATA concensus sequence. In this regard it is similar to other TATA elements in VZV promoters including those driving the expression of the gC, ORF 61, and IE62 genes [10, 22, 23, 28]. The significance of these atypical TATA elements is not known and the characterization of numerous other VZV promoters will be required in order to determine if a significant number of viral genes utilize such elements.

gI Promoter

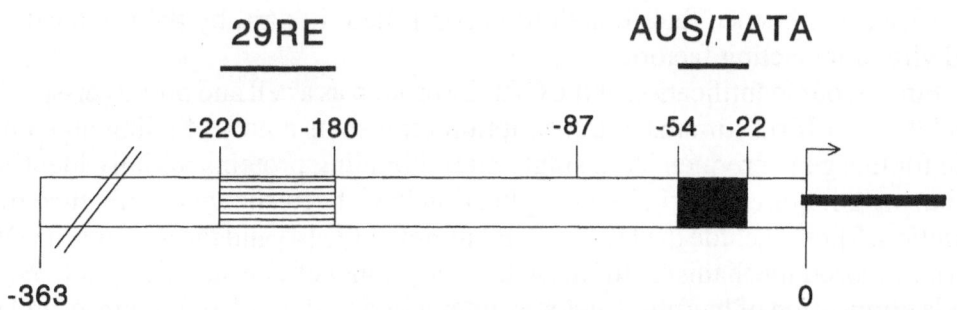

Fig. 3. Schematic diagram of the gI promoter indicating the positions of the 29RE and AUS/TATA sequences relative to the start of transcription

The AUS element contains overlapping binding sites for three cellular transcription factors including AP1, Sp1, and USF. The Sp1 and USF sites, 5′-ACGCCC-3′ and 5′-GTCACG-3′ are atypical when compared to the consensus binding sites for these factors [21, 27, 31, 39, 49] raising the possibility that they are not involved in expression of the gI promoter. Recent evidence from this laboratory, however, indicates that Sp1 and USF in extracts from both MeWo and T cells bind to the AUS sequence and that this binding is not dependent on the presence of viral proteins including IE 62. Furthermore, both deletion and site specific mutations which alter these sites result in an almost total loss of expression. In contrast, the AP-1 site appears to be dispensable since it is not bound by AP-1 and mutations at its 5′-end do not alter expression (H. He and W. Ruyechan, unpubl. observations).

USF binds to its recognition site and interacts directly with elements of the general transcription factor TFIID [48, 49]. USF has been shown to be involved in the activation of the bidirectional promoter controlling the expression of the VZV DNA polymerase (ORF 28) and major single-strand DNA binding protein (ORF 29) genes [34]. USF cooperates with the IE62 protein most likely via a direct physical interaction to activate expression of both genes. A recent analysis has shown that USF plays a critical role in the IE62 stimulated expression of the VZV ORF 4 promoter [35]. USF binding has also been reported at the promoter of the VZV ORF 10 gene [16]. VZV gI therefore is the fourth VZV gene whose expression is regulated, in part by USF suggesting that this ubiquitous cellular factor plays a major role in the expression of the VZV genome.

Sp1 is also a ubiquitous cellular transcription factor which binds to a variety of GC-rich sequences and interacts with TATA binding protein and other components of the general transcription factor TFIID [27]. The role of Sp1 in activation of VZV transcription has not been widely investigated. However, a recent report by Rahaus and Wolff [45] identified two closely spaced Sp1 binding sites upstream of the transcriptional start site of the VZV gE gene. One of these sites, 5′-GGGCGG-3′, fits the concensus Sp1 recognition sequence [27, 31]. The other, 5′-ACGCCC-3′ is identical to that which we have identified in the gI promoter. While the significance of the presence of the two atypical Sp1 sites in the gI and gE promoters is not obvious, it is intriguing to note that expression of these two glycoproteins (which form a heteromeric complex and whose relative abundances appear to significantly influence their function) is controlled, in part, by the same cellular and viral trans-acting factors.

Finally, our identification of the ORF 29 protein as a cell and promoter-specific modulator of IE62 stimulated transcription represents a novel finding and a new role for this gene product. Viral single-strand binding proteins initially identified as having a required role in DNA replication have been shown to influence transcription. These include the HSV-1 ICP8 protein [13, 14] and the adenovirus DBP [54]. Proposed mechanisms for these transcriptional effects are divergent and include elimination of hairpin structures, interaction with RNA polymerase, and the alteration of the binding affinities of transcription factors. We have shown that the ORF 29 protein's transcriptional modulatory activity requires the presence of a specific sequence and is significant only in T cells. These two observations suggest the existence of a repressor activity specific to T cells which can be overcome by the ORF 29 protein or, conversely, the absence of an activator for which the ORF 29 protein can somehow substitute. These possibilities are currently under investigation.

In summary, we have identified several cis-acting elements within the VZV gI promoter which are either required in the context of IE62 activated transcription or involved in cell type specific expression. We have also shown that the VZV ORF 29 protein is involved in the expression of the gI promoter along with the VZV IE62, ORF 4, and ORF 61 gene products and that its mechanism of action appears to be novel. Data accumulated from several laboratories indicate that the activity of these viral proteins is dependent on their relative abundance and on the presence or absence of cellular factors. Future experiments will be aimed at identification of the specific subsets of these viral and cellular proteins which are important for regulation of expression of VZV promoters during various phases of the viral life cycle.

Acknowledgements

The work from the authors' laboratories presented here was supported by grants AI18449 and AI36884 from the National Institute of Allergy and Infectious Diseases.

References

1. Boucaud D (1999) The varicella zoster virus open reading frame 29 protein: its role as a modulator of viral transcription. Ph.D. Thesis, SUNY at Buffalo, Buffalo, NY
2. Boucaud D, Yoshitake H, Hay J, Ruyechan WT (1998) The VZV ORF 29 protein acts as a modulator of a late VZV gene promoter. J Infect Dis 178 [Suppl 1]: 34–38
3. Croen KD, Ostrove JM, Dragovic LJ, Straus SE (1988) Patterns of gene expression and sites of latency in human nerve ganglia are different for varicella zoster and herpes simplex viruses. Proc Natl Acad Sci USA 85: 9773–9777
4. Cohen JI, Nguyen H (1997) Varicella zoster virus glycoprotein I is essential for growth of Virus in Vero cells. J Virol 71: 6913–6920
5. Cohen JI, Straus SE (1996) Varicella zoster virus and its replication. In: Fields BN, Knipe DM, Howley PM, Chanock RM, Melnick JL, Monath TP, Roizman B (eds) Fields virology. Lippincott-Raven, Philadelphia, pp 2525–2545
6. Cohrs RJ, Barbour M, Gilden DH (1996) Varicella zoster virus (VZV) transcription during latency in human ganglia: detection of transcripts mapping to genes 21, 29, 62, and 63 in a cDNA library enriched for VZV RNA. J Virol 70: 2789–2796
7. Cohrs RJ, Barbour MB, Mahalingam R, Wellish M, Gilden DH (1995) Varicella zoster virus (VZV) transcription during latency in human ganglia: prevalence of VZV gene 21 transcripts in latently infected human ganglia. J Virol 69: 2674–2678
8. Debrus S, Sadzot-Delvaux C, Nikkels AF, Piette J, Rentier B (1995) Varicella zoster virus gene 63 encodes an immediate early protein that is abundantly expressed during latency. J Virol 69: 3240–3245
9. Defechereux P, Melen L, Baudoux L, Merville-Louis M-P, Rentier B, Piette J (1993) Characterization of the regulatory functions of the varicella zoster virus open reading frame 4 gene product. J Virol 67: 4379–4385

10. Disney GH, McKee TA, Preston CM, Everett RD (1990) The product of varicella zoster gene 62 autoregulates its own promoter. J Gen Virol 71: 2999–3003

11. Davison AJ, Scott JE (1986) The complete DNA sequence varicella zoster virus. J Gen Virol 67: 2237–2242

12. Folks T, Benn S, Rabson A, Theodore T, Hoggan MD, Martin M, Lightfoote M, Sell K (1985) Characterization of a continuous T-cell line susceptible to the cytopathic effects of the acquired immunodeficiency syndrome (AIDS)-associated retrovirus. Proc Natl Acad Sci USA 82: 4539–4543

13. Gao M, Knipe DM (1991) Potential role for herpes simplex virus ICP8 DNA replication protein in stimulation of late gene expression. J Virol 65: 2666–2675

14. Godowski PJ, Knipe DM (1983) Mutations in the major DNA-binding protein of herpes simplex virus type 1 result in increased levels of viral gene expression. J Virol 47: 478–486

15. Graham FL, Van der Eb AJ (1973) A new technique for the assay of infectivity of human adenovirus 5 DNA. Virology 52: 456–467

16. Greaves RF, O'Hare P (1991) Sequence, function, and regulation of the Vmw65 gene of herpes simplex type 2. J Virol 65: 6705–6713

17. Greene LA, Aletta JM, Rukenstein A, Green SH (1987) PC12 pheochromocytoma cells: culture, nerve growth factor treatment, and experimental exploitation. Methods Enzymol 147: 207–216

18. Grose C (1990) Glycoproteins encoded by varicella zoster virus: biosynthesis, phosphorylation, and intracellular trafficking. Ann Rev Microbiol 44: 59–80

19. Hay J, Ruyechan WT (1994) Varicella zoster virus – a different kind of herpesvirus latency? Semin Virol 5: 241–247

20. Jackers P, Defechereux P, Baudoux L, Lambert C, Massaer M, Merville-Louis MP, Rentier B, Piette J (1992) Characterizations of regulatory functions of the varicella zoster virus gene-63 encoded protein. J Virol 66: 3899–3903

21. Janson L, Bark C, Pettersson U (1987) Identification of proteins interacting with the enhancer of human U2 small nuclear RNA genes. Nucleic Acids Res 15: 4997–5016

22. Kantakamalakul W (1994) Varicella zoster virus promoter sequences. PhD Thesis, University of the Health Sciences, Bethesda

23. Kantakamalakul W, Ruyechan WT, Hay J (1995) Analysis of varicella zoster virus Promoter sequences. Neurology 45 [Suppl 8]: 28–29

24. Kimura H, Straus SE, Williams RK (1997) Varicella zoster virus glycoproteins E and I expressed in insect cells form a heterodimer that requires the N-terminal of glycoprotein I. Virology 233: 382–391

25. Kinchington PR, Inchuaspe G, Subak-Sharpe JH, Robey F, Hay J, Ruyechan WT (1988) Identification and characterization of a varicella zoster virus DNA-binding protein by using antisera against a predicted synthetic oligopeptide. J Virol 62: 802–809

26. Kost RG, Kupinsky H, Straus SE (1995) Varicella-zoster virus gene 63: transcript mapping and regulatory activity. Virology 209: 218–224

27. Lania L, Majello B, De Luca P (1997) Transcriptional regulation by the Sp family proteins. Int J Biochem Cell Biol 29: 1313–1323

28. Ling PD, Kinchington PR, Ruyechan WT, Hay J (1990) A detailed transcriptional analysis of varicella zoster virus gene 14 (glycoprotein V). Virology 184: 625–635

29. Ling P, Kinchington PR, Ruyechan WT, Hay J (1992) Transcription from varicella-zoster virus gene 67 (glycoprotein IV). J Virol 66: 3690–3698

30. Litwin V, Jackson W, Grose C (1992) Receptor properties of two varicella zoster virus glycoproteins gpI and gpIV homologous to herpes simplex virus gE and gI. J Virol 66: 3643–3651

31. Locker J, Buzard G (1990) A dictionary of transcription control sequences. J DNA Seq Map 1: 3–11
32. Lungu O, Panagiotidis CA, Annunziato PA, Gershon AA, Silverstein SJ (1998) Aberrant Intracellular localization of varicella zoster virus regulatory proteins during latency. Proc Natl Acad Sci USA 93: 2122–2124
33. Mallory S, Sommer M, Arvin AM (1997) Mutational analysis of the role of glycoprotein I in varicella zoster virus replication and its effects on glycoprotein E conformation and trafficking. J Virol 71: 8279–8288
34. Meier JL, Luo X, Sawadogo M, Straus SE (1994) The cellular transcription factor USF cooperates with varicella-zoster virus immediate early protein 62 to symmetrically activate a bidirectional promoter. Mol Cell Biol 14: 6896–6906
35. Michael EJ, Kuck KM, Kinchington PR (1998) Anatomy of the varicella-zoster virus open reading frame 4 promoter. J Infect Dis 178 [Suppl 1]: 27–33
36. Moriuchi H, Moriuchi M, Straus SE, Cohen JI (1993) Varicella-zoster virus open reading frame 10 protein, the herpes simplex virus VP16 homolog, transactivates herpesvirus immediate-early gene promoters. J Virol 67: 2739–2746
37. Moriuchi H, Moriuchi M, Straus SE, Cohen JI (1993) Varicella-zoster virus (VZV) open reading frame 61 protein transactivates VZV gene promoters and enhances the infectivity of VZV DNA. J Virol 67: 4290–4295
38. Olson JK, Grose C (1998) Complex formation facilitates endocytosis of the varicella zoster GE:gI receptor. J Virol 72: 1542–1551
39. Osborne TF, Gil G, Brown MS, Kowal RC, Goldstein JL (1987) Identification of promoter elements required for in vitro transcription of hamster 3-hydroxymethylglutaryl coenzyme A reductase gene. Proc Natl Acad Sci USA 84: 3614–3618
40. Ostrove JM, Reinhold W, Fan C-M, Zorn S, Hay J, Straus SE (1985) Transcription mapping of the varicella zoster virus genome. J Virol 56: 600–606
41. Perera LP, Kaushal S, Kinchington PR, Mosca JD, Hayward GS, Straus SE (1994) Varicella-Zoster virus open reading frame (ORF) 4 encodes a transcriptional activator that is functionally distinct from that of herpes simplex virus homolog ICP27. J Virol 68: 2468–2477
42. Perera LP, Mosca JD, Ruyechan WT, Hay J (1992) Regulation of varicella zoster gene expression in human T lymphocytes. J Virol 66: 5298–5304
43. Perera LP, Mosca JD, Zadeghi-Zadeh M, Ruyechan WT, Hay J (1992) The varicella zoster virus immediate early protein IE62 can positively regulate its cognate promoter. Virology 191: 346–354
44. Perera LP, Mosca JD, Ruyechan WT, Hayward GS, Straus SE, Hay J (1993) A major transactivator of varicella zoster virus, the immediate early protein, IE62, contains a potent – terminal activator domain. J Virol 67: 4474–4483
45. Rahaus M, Wolff MH (1999) Influence of different cellular transcription factors on the regulation of varicella-zoster virus glycoproteins E (gE) and I (gI) UTR's activity. Virus Res 62 [SI 1]: 77–88
46. Roberts CR, Weir AC, Straus SE, Hay J, Ruyechan WT (1985) DNA-binding proteins present in varicella zoster virus infected cells. J Virol 55: 45–53
47. Ruyechan WT, Hay J (1999) Varicella zoster virus: molecular biology. In: Webster RG, Granoff A (eds) Encyclopedia of virology, 2nd ed. Academic Press, London, pp 1878–1884
48. Sawadogo M (1988) Multiple forms of the human gene specific transcription factor USF II. DNA binding properties and transcriptional activity of the purified HeLa USF. J Biol Chem 263: 11994–12001
49. Sawadogo M, Roeder RG (1985) Interaction of a gene-specific transcription factor

with the adenovirus major late promoter upstream of the TATA box region. Cell 43: 165–175

50. Tyler JK, Everett RD (1993) The DNA binding domain of the varicella zoster virus gene 62 protein interacts with multiple sequences which are similar to the binding site of the related protein of herpes simplex virus type 1. Nucleic Acids Res 21: 513–522

51. Wu C-L, Wilcox KW (1991) The conserved DNA-binding domains encoded by the herpes simplex virus type 1 ICP4, pseudorabies virus IE180, and varicella-zoster virus ORF62 genes recognize similar sites in the corresponding promoters. J Virol 65: 1149–1159

52. Ye M, Duus KM, Peng J, Price DH, Grose C (1999) Varicella-zoster virus Fc receptor component gI is phosphorylated on its endodomain by a cyclin-dependent kinase. J Virol 73: 1320–1330

53. Wietstock SM, Holmes AM, Ruyechan WT (1988) Identification and characterization of a DNA primase activity present in herpes simplex virus infected cells. J Virol 62: 1038–1045

54. Zijderveld DC, d'Adda di Fagnana F, Giacca M, Timmers HT, van der Vliet PC (1994) Stimulation of the adenovirus major late promoter in vitro by transcription factor USF is enhanced by the adenovirus DNA binding protein. J Virol 68: 8288–8295

Authors' address: Dr. W. T. Ruyechan, Department of Microbiology State University of New York at Buffalo, Buffalo, NY 14214, U.S.A.

Interactions among structural proteins of varicella zoster virus

M. Spengler[1,*], N. Niesen[1], C. Grose[2], W. T. Ruyechan[1], and J. Hay[1]

[1]Department of Microbiology and Witebsky Center for Microbial Pathogenesis
and Immunology State University of New York at Buffalo,
Buffalo, New York, U.S.A.
[2]Department of Pediatrics, University of Iowa, Iowa City, Iowa, U.S.A.

Summary. Varicella zoster virus tegument components include the regulatory proteins IE4, IE62, IE63 and the ORF10 protein, a protein kinase (ORF47) and an abundant protein encoded in ORF9 which is the homolog of HSV VP22. The kinase is able to phosphorylate IE62 and the ORF9 protein specifically in viral particles. We show that interactions among these proteins are, at least in part, dependent on the presence or absence of phosphate groups and we suggest models for tegument formation and for its dissolution in the infected cell.

Introduction

Varicella zoster virus (VZV) is a common human alphaherpesvirus that causes two main disease problems – chickenpox, a result of primary infection, and shingles, caused by the reactivation of latent virus.

The study of VZV has been hampered by the fact that the virus is a strict human pathogen and grows only to low titres in cell culture. Two milestones have substantially assisted the development of molecular genetics of VZV over the past ten years – the sequencing of the viral genome and the production of infectious virus from cosmids [1, 3]. Another recent advance in our understanding of VZV disease has been the widespread use of the Oka VZV vaccine [6, 15], although it is still unclear what constitutes the attenuation genotype or phenotype for this vaccine virus.

The VZV particle was first viewed by light microscopy in vesicle fluids recovered from patients and electron microscopic studies showed the viruses of chickenpox and shingles to be indistinguishable [5, 16]. This was subsequently confirmed by molecular techniques [14]. The virion is made up of four parts – the core, the capsid, the tegument and the envelope. Of these, the tegument is the

*Present address: Roswell Park Cancer Institute, Elm and Carlton Streets, Buffalo, NY 14263, U.S.A.

least understood but likely contains the majority of the virion proteins. Data from other herpesviruses suggest at least some of the tegument proteins are present in an ordered structure within the virion [18].

It has been possible to identify some of the VZV tegument proteins, and these include IE62, IE63, IE4, the ORF10 protein and the ORF47 protein kinase [7, 8]. All of these appear to play some role in viral gene regulation, although the actual role of the kinase remains speculative. IE62 is a major controlling protein in the cascade of expression from VZV genes. It is a 1310 aa nuclear phosphoprotein with, among other properties, a DNA-binding region and an N-terminal activation domain, similar in structure and strength to that of the herpes simplex virus (HSV) transactivator, VP16 [2, 7]. The other proteins listed above, while in certain circumstances having a direct role in gene regulation, may well be most important in the VZV growth cycle for their ability to assist IE62.

In this paper, we deal with potential interactions among IE62, IE4 and IE63 and the ORF9 protein; we also propose a model for the assembly and disassembly of the VZV particle tegument structure.

Materials and methods

Cells and viruses

MeWo cells, a human melanoma line, kindly supplied by Dr. C. Grose, University of Iowa, was our choice for growth of VZV. They were grown in Eagle's MEM with 2 mM glutamine, 100 U penicillin and 100 μg streptomycin/ml and 10% fetal bovine serum. Approximately 10% of the contents of a 150 cm^2 flask at 80% CPE due to the VZV Oka strain were used to infect further MeWo cells.

IE62

Cloned VZV IE62 ORF sequences were expanded in pGEM and transferred to the baculovirus shuttle vector pBacPak9 for recombination with wild-type virus. Recombinants were triply plaque-purified and tested for expression of authentic IE62. Recombinant virus-infected Sf21 insect cell nuclear extracts were subjected to ion exchange chromatography on Q sepharose followed by Sp sepharose. The final product was about 90% pure based on stained gel patterns.

IE4

The plasmid pVFsp4 with the N-terminal 44 kb of the VZV genome was digested with EcoR1 and Nco1 to yield a 4 kb fragment, which was further digested with Sal1 and ligated into pMal2c. Expressed proteins were induced by IPTG, released by sonication and eluted from amylose resin with 10 mM maltose.

IE63

IE63 was expressed as a glutathione-S-transferase fusion protein using the pGEX system, as described by Stevenson et al. [13].

The ORF9 protein

The ORF9 protein was expressed as a maltose fusion protein construct. Sequences bearing the ORF9 were generated from pVFsp4 using PCR and inserted into pMal2c that had been cut with Stu1 and Sal1. Expressed proteins were treated as above.

Antibody production

Polyclonal rabbit antibodies were produced by separation of the antigen in question on PAGE, mincing of the gel fragment in PBS and complete Freund's adjuvant, followed by subcutaneous injection of New Zealand White rabbits. Booster doses were given by the same route, as required, using incomplete Freund's. Development of antibodies was followed by ELISA. Antibodies to both IE62 and the ORF9 protein were made in this way. Rabbit polyclonal antibodies to IE4 and IE63 were kindly supplied by Dr. P. R. Kinchington, University of Pittsburgh.

Monoclonal antibody to IE62 was generated from the H6 hybridoma (a gift of Dr. A. Arvin, Stanford University) and was purified using E-Z-Step (Pharmacia, Inc.).

SDS-polyacrylamide gel electrophoresis (SDS PAGE)

This was carried out using the minigel system of Promega Inc. (Protocols and Applications Guide, 1991).

Western blotting

Proteins separated on SDS PAGE were transferred to nitrocellulose sheets in a BioRad transfer unit [17]. After washing with 2% dried milk in PBS for 1 h, the membrane was incubated with primary antiserum at dilutions which varied with the serum used, then secondary antiserum, and developed using the ECL system of Amersham Life Sciences.

Farwestern blotting

Proteins were resolved on SDS PAGE and electroblotted to PDVF membranes (Schleicher and Schuell Inc.). Following renaturation overnight in the presence of 2% dried milk, the probe protein was added for an hour. The blots were developed using the ECL system.

Results

IE62 and IE4

There are substantial data to suggest that VZV IE4 may interact physically with VZV IE62. For example, it has been known for some time that these two proteins can act in a synergistic sense to upregulate transcription from a variety of viral promoters in in vitro assays with reporter constructs [10]. Also, in more recent experiments, we have confirmed that IE62 is capable of targeting IE4 to the nucleus of cells. The simplest explanation for these findings would be that IE62 and IE4 are able to interact physically.

We have carried out experiments to test for possible physical interaction by initially treating VZV Oka – infected cells with anti-IE62 or anti-IE4 antibodies and collecting the immune precipitates. When these were resolved on PAGE, the IE62 precipitate contained IE4, and the IE4 precipitate contained IE62, suggesting that, in the VZV-infected cell, these two proteins exist in a complex and could have direct affinity for each other. To assess their ability to interact directly, both IE62 and IE4 were expressed in vitro in a variety of systems (bacterial and insect) and a number of direct interaction assays were carried out. Figure 1 gives the results of one of these, an antibody pull-down assay. In this assay, maltose binding protein (MBP) or MBP fused to IE4 were mixed with IE62 purified

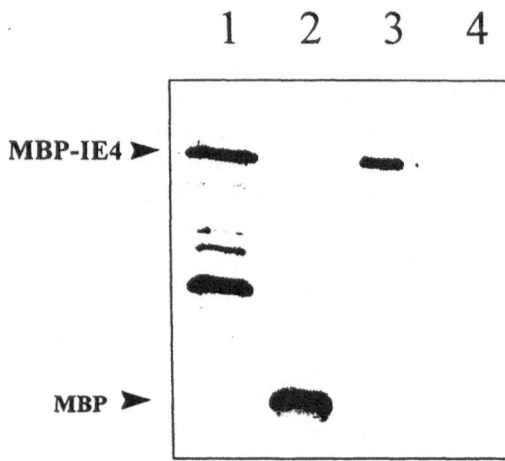

Fig. 1. Anti-IE62 antibody coimmunopre-
cipitates IE62 and IE4. Purified MBP-IE4
or MBP were mixed with IE62, protein
G sepharose and a monclonal anti-IE62
antibody. The washed precipitates were
resolved by SDS PAGE and blotted. A western assay was carried out using anti-MBP
antibody *1* MBP-IE4; *2* MBP; *3* MBP-IE4,
IE62 and anti-IE62 antibody; *4* MBP, IE62
and anti-IE62 antibody

from baculovirus-infected cells, a monoclonal antibody to IE62 and protein G
sepharose. The washed immune precipitates were resolved on PAGE and western
blotted with antibody against MBP. In Fig. 1, lane 1 and 2 are the MBP and MBP-
IE4 proteins as reference, while lanes 3 and 4 are precipitates with MBP-IE4 and
MBP respectively. Clearly, anti-IE62 antibody can specifically precipitate IE4,
suggesting a direct interaction between the two proteins. When the experiment
was rerun with anti-MBP antibody as the primary precipitant, again IE62 and IE4
specifically co-precipitated. Several other assays were performed to confirm this
interaction including an ELISA, and a glutathione-S-transferase (GST) capture
assay; all assays showed specific interaction between the two proteins.

IE62 and IE63

Using arguments similar to those used for IE4, based primarily on the ability of
IE63 to modify the transactivation functions of IE62 ([10]; Mathieson, Ruyechan
and Hay, unpubl. observations), we set out to examine possible interactions between IE62 and IE63. In a similar experiment to that outlined above for IE4, we
precipitated VZV-infected cell extracts with antibodies and looked for the presence of IE62 and/or IE63 in the extracts. Our data showed that IE62 and IE63
are present in a complex in VZV-infected cells. To find out if these two proteins
interact directly, we expressed IE63 in vitro and carried out several binding assays. One of these is shown in Fig. 2. Here, GST or a GST-IE63 fusion protein was
attached to sepharose beads; these were treated with purified baculovirus-derived
IE62 and washed. The washed beads were analyzed by SDS-PAGE, followed
by western blotting using a rabbit polyclonal anti-IE62 antibody. IE62 binds to
GST-IE63 but not to GST, suggesting a direct interaction. Additional assays, including ELISA with intact and fragmented IE62 and IE63 confirmed that these
two proteins can interact directly. Other assays with these proteins fail to show
an interaction in vitro between IE63 and IE4, although they are both likely to be
part of the VZV-infected cell complex with IE62.

VZV   GST   GST-IE63

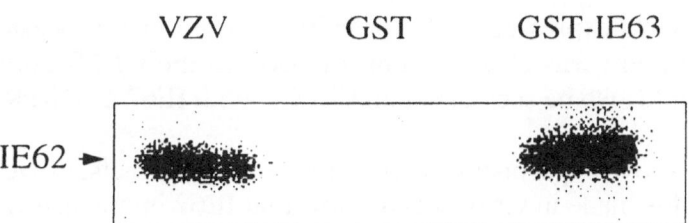

IE62 →

Fig. 2. GST-IE63 protein captures IE62. IE62 purified from insect cells was reacted with GST or GST fused with IE63 bound to sepharose. Proteins eluted with glutathione were run on SDS PAGE and blotted with anti-IE62 antibody

IE62 and the ORF9 protein

One of the properties of IE62 that emerged soon after its discovery was that it had features in common with the HSV transactivator, VP16; in particular, the acidic activation domains of both proteins were strikingly similar [1, 11]. The activation domain of VP16 is involved in binding to VP22 and, when the complex is formed, the proteins appear to relocate to a structure suspected to be the viral tegument [4]. The VZV equivalent of HSV VP22 is the ORF9 protein, an unremarkable polypeptide of about 30 kDa, according to the sequence data of Davison and Scott [3]. We cloned and expressed the VZV ORF9 in a variety of systems, and developed polyclonal rabbit antisera to the cloned proteins. Using these reagents, we were able to show that the ORF9 protein appears on PAGE as a heterogeneous set of polypeptides, ranging in apparent m.wt. from about 35 kDa to 42 kDa; it is also a prominent viral structural protein. In order to investigate whether the ORF9 protein had binding properties for IE62, we carried out a series of assays analogous to those described above for IE4 and IE63. To summarize, IE62 and ORF9 protein were able to be coprecipitated from VZV-infected cell extracts

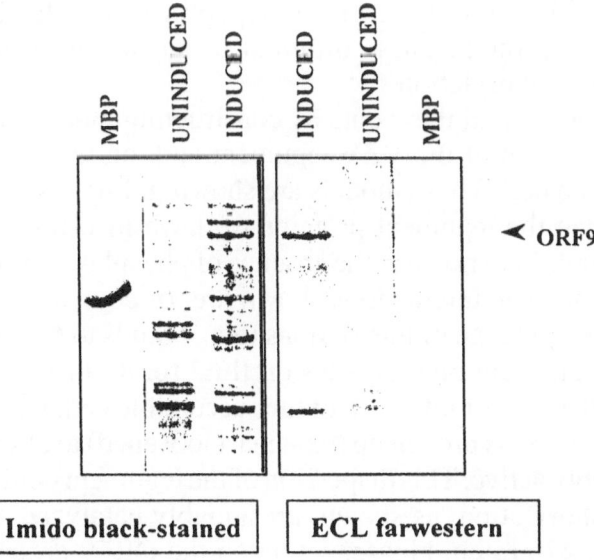

◄ ORF9

Imido black-stained   ECL farwestern

Fig. 3. IE62 has affinity for the VZV ORF9 protein. *E. coli* bearing the plasmid encoding MBP-ORF9 were uninduced or induced with IPTG and proteins resolved on SDS PAGE. The gel was stained with amido black (left panel) or subjected to farwestern blotting with IE62 and anti-IE62 antibody (right panel)

and, when in vitro-expressed ORF9 protein was used in a far western assay with IE62 (Fig. 3) there was clear binding of IE62 to the ORF9 polypeptide. This interaction could also be seen using an ELISA, with IE62 or ORF9 protein as the bait.

Thus, the above experiments, using infected cell extracts, as well as partially-purified proteins made in vitro clearly show that IE62 interacts directly with IE4, IE63 and the ORF9 protein.

Discussion

Our initial interest in interactions among these VZV proteins stemmed from our curiosity about viral gene expression and its control. In that context, one possibility is that the differential relationships between IE62 and the various promoters with which it interacts during infection are controlled by the presence or absence of auxiliary activators/repressors such as IE4, IE63 or the ORF9 protein, although recent data from Perera [12] infers that IE62 alone may be able to discriminate among promoters based simply on the TATA box sequence.

As the data emerged from these studies, however, it became clear that, since all of these proteins are tegument proteins, their ability to interact may also have importance for tegument structure. In the course of these studies, we also realized that interactions between IE4 and IE62 and between the ORF9 protein and IE62 were influenced by phosphorylation. Specifically, phosphorylation of IE62 at T_{250} completely abrogated binding of IE4. Similarly, phosphorylation of the ORF9 protein interfered with the ability of IE62 to bind to it. In an earlier study [7], we had shown that a protein kinase in VZV particles (the ORF47 protein) was able to phosphorylate in situ polypeptides of 175 kDa, 42 kDa and 37 kDa. This was shown by purifying VZV virions and treating them with low concentrations of NP40 and ^{32}P-ATP. We identified the largest polypeptide as IE62 but only when the reagents recently became available to identify the ORF9 protein were we able to show that both the 37 kDa and the 42 kDa polypeptides were likely to be different forms of the ORF9 protein. These different forms appear to be distinguished by their levels of phosphorylation.

We have attempted to make sense of all these data by constructing models of, on the one hand, steps in the formation of the VZV tegument and, on the other hand, of its dispersal in the infected cell. These models are shown in Fig. 4A and B. In Fig. 4A, we hypothesize that the tegument proteins we have investigated [ORFs 4, 9, 47, 62, 63] form a complex as shown in the absence of phosphorylation or at low levels of phosphorylation. The interactions shown are, of course, only schematic but reflect our current knowledge of interactions (e.g. 4 binds to 62 and 9) and our recent data that show different binding sites on IE62 for 4, 9 and 63. Figure 4B shows the situation at the time of infection of the susceptible cell, when the tegument complex, we hypothesize, is broken up (or at least loosened) to allow its component proteins to be suitably active. This dispersion of the tegument could, in principle, be driven by phosphorylation, as shown, presumably catalyzed, at least in part, by the resident ORF47 protein kinase, acting on the ORF9 protein

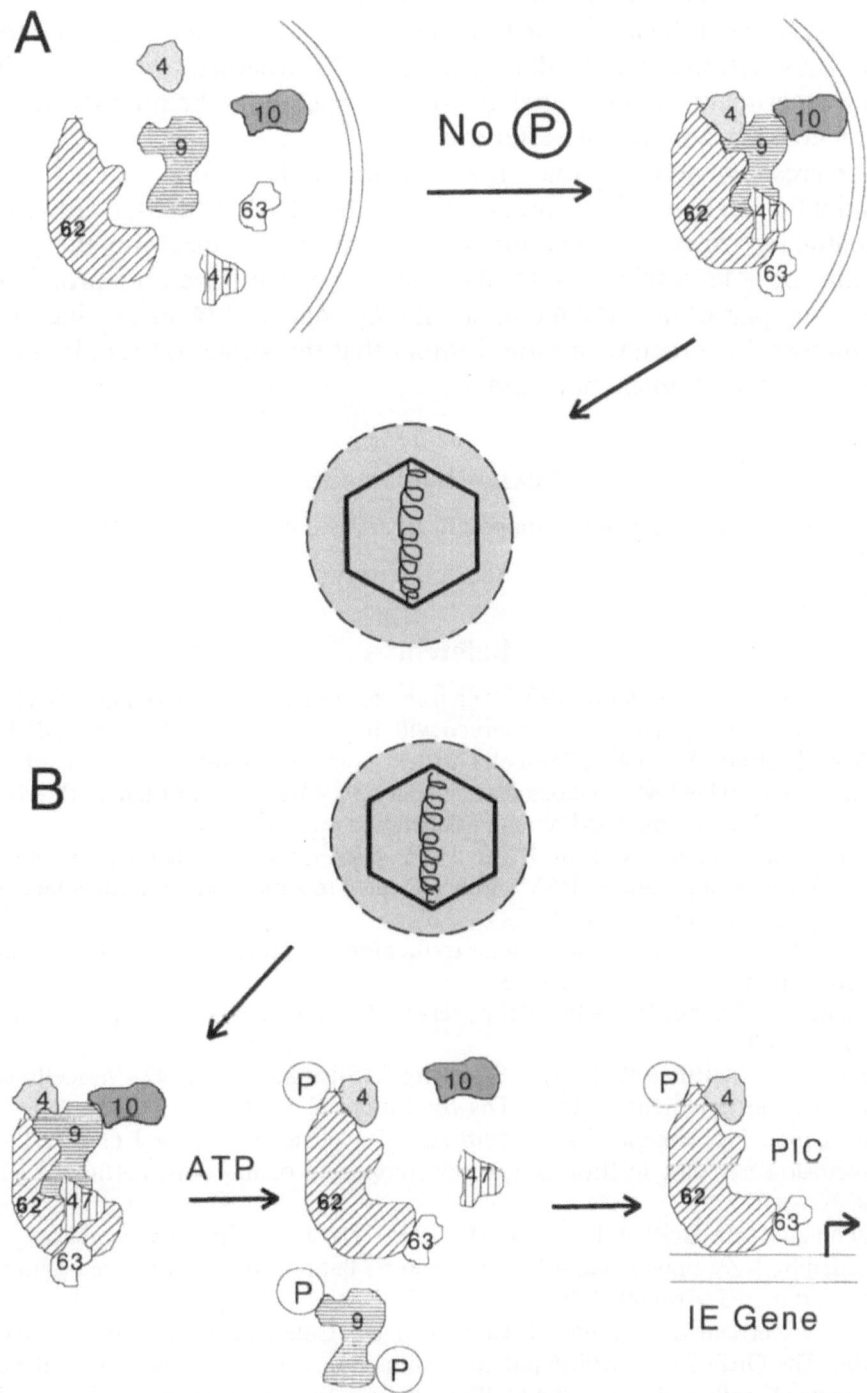

Fig. 4. Models for tegument construction and dissolution. In **A**, we show a putative scheme for assembly of VZV tegument; in **B**, the possible mode of release of viral proteins from the tegument after infection, under the influence of phosphorylation. *PIC* Preinitiation complex

and, perhaps, on IE62. We have added the ORF10 protein to the models in an arbitrary way; we know it is present in tegument but we did not investigate its relationships with the other tegument proteins.

At least some of the above ideas are testable, and we have recently subjected purified VZV virions to mild detergent and ATP treatment, observing in preliminary experiments that the nucleocaspsids appeared to be released from their tegument coating more readily after treatment.

An interesting endnote is that the Arvin lab has shown that the ORF47 kinase is essential for viral growth in human tissue in the SCID-hu model. However, it is not essential for growth in cell culture [9]. This raises the possibility that a cellular kinase may be able to substitute for the viral enzyme and that, if a protein kinase is a necessary part of the viral tegument, it may conceivably also be incorporated into tegument. Such behavior would imply that the structural requirements for tegument incorporation are not rigid.

Acknowledgement

The authors gratefully acknowledge support to JH and WTR from the USPHS (AI36884 and AI18849).

References

1. Cohen JI, Heffel D, Seidel K (1993) The transcriptional activation domain of VZV open reading frame 62 protein is not conserved with its HSV homolog. J Virol 66: 4226–4251
2. Cohen JI, Straus SE (1996) Varicella zoster virus and its replication. In: Fields BN, Knipe DM, Howley PM, Chanock RM, Melnick JL, Monath TP, Roizman B (eds) Fields Virology, 3rd ed. Lippincott-Raven, Philadelphia, pp 2525–2545
3. Elliot G, Mouzakitis G, O'Hare P (1995) VP16 interacts via its activation domain with VP22, a tegument protein of HSV and is relocated to a novel macromolecular assembly in coexpressing cells. J Virol 69: 7932–7941
4. Elliot G, O'Hare P (1997) Intercellular trafficking and protein delivery by a herpesvirus structural protein. Cell 88: 223–233
5. Farrant JL, O'Connor JL (1949) Elementary bodies of varicella and herpes zoster. Nature 163: 260–263
6. Gershon AA, LaRussa P, Hardy I, Steinberg S, Silverstein S (1992) Varicella vaccine: The American experience. J Infect Dis 166 [Suppl 1]: 63–68
7. Kinchington PR, Hougland JK, Arvin AM, Ruyechan WT, Hay J (1992) The VZV immediate early protein IE62 is a major component of the virus particle. J Virol 66: 359–366
8. Kinchington PR, Bookey D, Turse SE (1995) The transcriptional regulatory proteins encoded by VZV open reading frames 4 and 63 but not 61 are associated with purified virus particles. J Virol 69: 4272–4282
9. Moffat J, Zerboni L, Sommer M, Heineman TC, Cohen JI, Kaneshima H, Arvin AM (1998) The ORF47 and ORF66 putative protein kinases of VZV determine the tropism for human T cells and skin in the SCID-hu mouse. Proc Natl Acad Sci USA 95: 11969–11974
10. Perera LP, Mosca J, Ruyechan WT, Hay J (1992) Regulation of VZV gene expression in human T lymphocytes. J Virol 66: 5298–5204

11. Perera LP, Mosca J, Ruychean WT, Hayward GS, Straus SE, Hay J (1993) A major transactivator of VZV, the immediate early protein IE62, contains a potent N-terminal activation domain. J Virol 67: 4474–4483

12. Perera LP (2000) The TATA motif specifies the differential activation of minimal promoters by VZV immediate early regulatory protein IE62. J Biol Chem 275: 487–496

13. Stevenson D, Xue M, Hay J, Ruyechan WT (1996) Phosphorylation and nuclear localization of the VZV gene63 protein. J Virol 70: 658–662

14. Straus SE, Reinhold W, Smith HA (1984) Endonuclease analysis of viral DNAs from varicella and subsequent zoster infections in the same patient. N Eng J Med 311: 1362–1364

15. Takahashi M, Okuno Y, Otsuka T, Osama J, Takamizawa A, Sasaha T, Kubo T (1975) Development of a live attenuated varicella vaccine. Biken J 18: 25–33

16. Tournier P, Cathala F, Bernhard W (1957) Ultrastructure et developpement intracellulaire du virus de la varicelle observe au microscope electronique. Presse Med 65: 1229–1234

17. Towbin H, Gordan J (1984) Immunoblotting and dot immunoblotting – current status and outlook. J Immunol Methods 72: 313–315

18. Zhou ZH, Chen DH, Jakana J, Rixon F, Chiu W (1999) Visualization of tegument-capsid interactions and DNA in intact HSV type 1 capsids. J Virol 73: 3210–3218

Author's address: Dr. J. Hay, Department of Microbiology and Witebsky Center for Microbial Pathogenesis and Immunology, State University of New York at Buffalo, Buffalo, NY 14214, U.S.A.

The role of varicella zoster virus immediate-early proteins in latency and their potential use as components of vaccines

C. Sadzot-Delvaux and **B. Rentier**

Department of Microbiology, Fundamental Virology, Liège University,
Sart Tilman-Liège, Belgium

Summary. Varicella zoster virus immediate-early (IE) proteins are intracellular regulators of viral gene expression. Some of them (IE62 and IE63) are found in large amounts in infected cells but are also components of the virion tegument. Several IE and early genes are transcribed during latency, while late genes are not. Recently, we demonstrated the presence of protein IE 63 in dorsal root ganglia of persistently infected rats as well as in normal human ganglia; other IE proteins have been found since in human ganglia. Cell-mediated immunity (CMI) to IE 62 has been evidenced. We found both humoral immunity and CMI to IE 63 in immune adults. In elderly zoster-free individuals, CMI to IE 63 remained high. The differences in the CMI to IE 63 among young adults, elderly people and immunocompromized patients have to be analyzed according to their status relative to zoster, to determine whether the decrease in CMI, particularly to IE proteins, could be responsible for viral reactivation and for the onset of shingles. Hopefully, the waning of the CMI to VZV IE 63 and perhaps to other IE proteins could become a predictive marker for herpes zoster and reimmunization, not only with the vaccine strain, but also with purified IE proteins could help prevent zoster at old age.

Introduction

Varicella zoster virus (VZV) has long been considered as a virus whose functioning could be deduced from that of herpes simplex virus (HSV), the prototype of Alphaherpesviruses, which is easy to produce in vitro and for which experimental models are available. Indeed VZV shares many morphological and biological properties with HSV and all other alphaherpesviruses. The most interesting characteristic certainly is the fact that, following primary infection, the virus can reach sensory ganglia and remain latent in the peripheral nervous system for many years before being reactivated and producing a rash, usually restricted to a single dermatoma.

However, it is now obvious that the mechanisms involved in VZV latency and reactivation are not identical to and perhaps not even similar to those described for HSV. HSV-1 remains latent in neurons only, and its latency is characterized by the accumulation of antisense latency-associated transcripts (LATs) in the nucleus of infected cells. Although these transcripts are thought to contain at least one open reading frame, no corresponding protein has been detected so far and it is still not clear whether the LATs play a role in the induction or in the maintenance of latency, while they could be involved in viral reactivation [44]. These characteristics are shared by the other alphaherpesviruses such as pseudorabies virus, equine herpesvirus-1 or bovine herpesvirus-1 but not by VZV, whose genome lacks any homolog of the HSV LAT-encoding sequence.

In situ hybridization and Northern blot analysis have provided evidence for a restricted transcription of the VZV genome during latency: both α (ORFs 4, 62 and 63) and putative β genes (ORFs 29 and 21) are transcribed during latency while no γ gene is [9–11, 31, 43]. The viral proteins corresponding to these transcripts have now been detected in rat [12] and in latently infected human ganglia [28, 30]. Three IE proteins (IE 4, 62 and 63) and two E proteins (ORF21p and ORF29p) accumulate in latently infected cells while no late protein seems to be expressed.

Key roles of the immediate-early proteins in productive infection

The VZV replication cycle can be subdivided into three phases: (1) virus adsorption and entry, (2) viral gene transcription and translation and (3) viral assembly and egress. Once the virus has penetrated the cell by fusion of its envelope with the plasma membrane, involving interactions between viral glycoproteins and cell receptors, the nucleocapsid reaches a nuclear pore through which the viral genome is released into the nucleoplasm. Viral genes are transcribed in the nucleus and transcripts are translated in the cytoplasm in three successive phases. A first wave leads to the very early expression of ORFs 4, 61, 62 and 63, in the absence of de novo protein synthesis. The proteins produced, called immediate-early proteins (IE), migrate back into the nucleus and induce the expression of a second class of proteins called early (E), mainly the enzymes involved in viral DNA replication. The third and last wave of protein expression occurs after DNA replication and results in the synthesis of late (L) proteins that will constitute the viral particle. The capsids are assembled in the nucleus, wrapped around the newly replicated DNA before acquiring an envelope and being released.

In this highly controlled process, the expression of the IE proteins is certainly the most critical step since these proteins play a key role in the regulation of expression of most other viral genes. Moreover, except for ORF61p, these proteins are found associated with purified virus particles [23, 24] and could constitute important targets for the immune system.

IE 62

ORF62p, a 175 kDa protein encoded by ORF 62 and also by ORF 71, is present in large amounts in the viral tegument [24] and thus brought in by the incoming virus. It is expressed as a nuclear IE phosphoprotein, thus it is called IE 62 and it exerts an important regulatory function in VZV replication, as evidenced by transient expression experiments demonstrating that it is capable of transactivating genes of all classes and increasing viral DNA infectivity [21, 38, 41]. Moreover, IE 62 regulates its own promoter [5, 16]. However, these regulatory properties appear to be dose-dependent because IE 62 can positively or negatively regulate the expression of other genes, depending on its concentration.

However, IE 62 does probably not act on its own, but in synergy with other IE proteins or even with cell proteins. A direct interaction with IE 4 specifically modifies the intracellular localization of the latter by transporting it to the nucleus [14]. Based on these data, it seems reasonable to propose that IE 62, brought in at a low concentration by the incoming virus, initiates the IE phase by transactivating the expression of α genes. In a second step, IE 62 could participate in the activation of β genes and later of γ genes. Due to its autoactivating properties, its intracellular concentration increases progressively and its expression is thereby down-regulated.

IE 4

ORF4p, a 55 kDa protein, is also present in large amounts in the viral tegument [23] and produced very early on in infected cells [13]. Its localization is mostly cytoplasmic but thanks to a direct interaction with IE 62, IE 4 can be transported to the nucleus [13]. IE 4 transactivates promoters of all three classes of genes, either on its own or in synergy with IE 62 [15], but it has no demonstrable transrepressing activity [37]. Its transactivating properties could require the presence of other functional cellular proteins.

IE 63

ORF63p is a 45 kDa phosphoprotein encoded by genes 63 and 72. It is a component of the viral tegument, shown to be a true IE protein and it is strongly expressed during lytic infection in cell culture as well as in skin lesions [12, 23, 34]. Its localization is mostly nuclear, but it can also be detected in the cytoplasm of infected cells [12]. Even though IE 63 shows a slight repression of IE 4 transactivating properties on the IE 62 promoter [12], it lacks significant transactivating properties and its role at the very beginning of the replicative cycle has not yet been elucidated.

ORF61p

The phosphoprotein encoded by ORF 61 (ORF61p, 62–65 kDa) has been considered as an IE protein only on the basis on its homology with HSV-1 ICP-0, but an experimental demonstration of its IE nature has not yet been provided. Contrary to the three other IE proteins, ORF61p does not belong to the viral tegument [23]. It enhances infectivity of VZV and HSV-1 DNA as shown by transient expression experiments [32] and, depending on its cellular concentration and on host cell lines, it can either repress or transactivate the functions of IE 4 and IE 62 on other VZV gene promoters [8, 32, 33].

Latency

Besides their functional properties as regulatory proteins in productive infection, proteins IE 4, 62 and 63 appear to be of great interest because they are also expressed in large amounts during latency. An expression of viral proteins has never been described during alphaherpesvirus latency, but it has been demonstrated for IE 63, first in an animal model [12, 39] then in humans [30]. Later, other IE or E proteins have been detected in latently infected human cells [28]. Mostly intranuclear during productive infection according to their functions, these proteins are detected predominantly in the cytoplasm of latently infected cells but they become detectable in nuclei when the virus reactivates. The reasons for this modified distribution during latency are still not understood. It is also not clear whether cytoplasmic sequestration is the result of a failed process that normally carries them to the nucleus where they perform their regulatory tasks or whether their accumulation by itself inhibits replication. Cytoplasmic overloading of these proteins during latency suggests new hypotheses for the establishment and maintenance of VZV latency [27]: the virus enters the cell as in productive infection and the cycle initiates by the expression of some of the regulatory IE proteins (IE 4, 62 and 63) which probably migrate to the nucleus to exert their regulatory functions, but they are in such low quantities that they remain undetectable.

Even a low amount of IE proteins migrating to the nucleus allows the expression of E proteins encoded by ORFs 21 and 29 [28], which also accumulate in large amounts in the cytoplasm. It is not known whether other E proteins are expressed and whether viral DNA replication occurs at all, but so far, no L protein has been detected during latency.

The mechanisms involved in this process have yet to be identified, but it is tempting to conclude that cells in which the virus becomes latent lack the necessary elements to process IE proteins and to give them the conformation by which they become functional. Phosphorylation or other post-translational modifications are candidate impaired process mechanisms. It is also possible that some cellular components interact with viral proteins or with viral promoters that can interfere with the replication cycle and lead to replicative arrest. For instance, Patel and collaborators have demonstrated that the isoforms of the cellular transcription factor Oct-2 expressed specifically in neuronal cells can inhibit basal activity of the VZV IE 62 promoter in neuronal cells but not in other cell types, suggesting a cell type-specificity [35]. This mechanism could be of particular importance for the onset of VZV latency in sensory ganglia.

However, such explanations still need experimental confirmation and must take into account the observation that inhibition is reversible in yet undetermined conditions when reactivation occurs.

Another important difference between VZV and HSV latency is the nature of the cells in which the virus remains latent. HSV clearly resides in neurons only whereas the precise localization of persistent VZV is still debated: using *in situ* hybridization or *in situ* PCR, evidence of latency in neuronal cells [17, 20], in non neuronal cells [11, 31] or in both cell types [26, 29] has been shown. Animal models did not allow elucidation of this issue because in mice and rats the viral genome was detected in both cell types [39, 45].

Immunogenicity of the IE proteins

One of the parameters of obvious importance for the control of virus infection is host immunity. Clinical observations have shown that the frequency and severity of viral reactivations increase in patients whose cell-mediated immunity (CMI) is impaired because of age, pathological disorders or immunosuppressive treatments prior to transplantation [3, 6, 18].

Previous studies have shown that viral tegument proteins, including IE 62 and 63, and the major glycoproteins, are important targets for CMI to VZV. They elicit a long term humoral and cellular immune response after natural VZV infection [1, 4, 40]. T lymphocytes from most VZV naturally immune donors proliferate in vitro after stimulation with these proteins and they can lyse autologous target cells that express IE 62, 63, gC, gE, gG or gI.

The critical role of immune control has been suggested by Hope-Simpson as early as 1965 [19]: VZV primary infection appears to be limited by host "resistance" (immune response) that remains high for many years, with a slow decrease over time. This decrease could, however, be partly counterbalanced by frequent viral reactivations or contact with infected individuals, which would contribute to maintaining an efficient immunity until reaching a critical level under which the host resistance would be too low to control viral reactivation. This hypothesis, based only on clinical observations is still valid today even if it appears now that it is mostly the CMI that limits viral reactivation. Indeed, virus reactivates in spite of high anti-VZV antibody titers and the zoster episode is not correlated with hypogammaglobulinæmia.

It is tempting to think that the expression of viral proteins in latently infected cells constitutes another way to trigger the immune response and could thus contribute to maintain it at a protective level. In this context, it will be of interest to characterize the immune response to viral proteins expressed during latency, in the elderly and in immunocompromised patients who have a high incidence of herpes zoster. It is well known that the CMI is often impaired with age, as documented using in vitro PBMC stimulation by whole VZV antigens [18]. Such a study has not been performed with purified VZV proteins and in particular with proteins expressed in latently infected cells.

However, many questions are raised by this hypothesis of a role for the viral 'latency proteins' in maintaining a specific immune response. The nervous system is indeed protected from immune recognition by anatomical barriers and neurons lack classical MHC molecules at their surfaces . On the contrary, satellite cells surrounding neurons express MHC and they could play an important role in antigen presentation. It is thus critical to define clearly in which cells the virus remains quiescent. So far, during latency, IE 63 has been observed only in neuron cytoplasms. If neurons are the only cells expressing viral antigens, the mechanisms leading to the recognition of viral peptides must be clarified. It is possible that non-classical MHC proteins are being expressed at the cell surface in response to viral infection, as it has been suggested for HSV [36]. However, in HSV-infected cells, peptide presentation by MHC molecules appears to be impaired because of a viral protein that inhibits peptide processing [22, 42]. VZV has been shown to selectively downregulate cell-surface MHC class I expression on human fibroblast cells and to inhibit IFN-γ induction of MHC class II cell surface expression [2].

Enhancing the immune response to IE 63 may become an important strategy for preventing VZV reactivation from latency. If so, IE 63 could be a suitable candidate as an additive for a VZV vaccine to be given to ageing adults in order to boost their immunity and to prevent herpes zoster [7, 25].

Because recognition by the immune system of proteins expressed during latency, proteins such as IE 62 and 63, could play a role in the control of latency, it must be documented using animal models or by studies involving a larger number of donors, particularly donors with a high probability of reactivation of the virus.

Acknowledgements

We wish to thank Drs. M. Takahashi and K. Yamanishi, the VZV Research Foundation and the "Fonds de la Recherche Scientifique Médicale" of Belgium for their support.

References

1. Ahn K, Meyer TH, Uebel S, Sempe S, Djaballah H, Yang Y, Peterson PA, Fruh K, Tampe R (1996) Molecular mechanism and species specificity of TAP inhibition by herpes simplex virus ICP47. EMBO J 15: 3247–3255
2. Arvin AM, Kinney-Thomas E, Shriver K, Grose C, Koropchak CM, Scranton E, Wittek AE, Diaz PS (1986) Immunity to varicella-zoster viral glycoproteins, gp I (gp 90/58) and gp III (gp 118), and to a nonglycosylated protein, p 170. J Immunol 137: 1346–1351
3. Arvin AM, Pollard RB, Ramussen LE, Merigan TC (1980) Cellular and humoral immunity in lymphoma patients. J Clin Invest 65: 869
4. Arvin AM, Sharp M, Smith S, Koropchak CM, Diaz PS, Kinchongton P, Ruyechan W (1991) Equivalent recognition of a varicella-zoster virus immediate early protein (IE62) and glycoprotein I by cytotoxic T lymphocytes of either CD4+ or CD8+ phenotype. J Immunol 146: 257–264
5. Baudoux L, Defechereux P, Schoonbroodt S, Merville MP, Rentier B, Piette J (1995) Mutational analysis of varicella-zoster virus major immediate-early protein IE62. Nucleic Acids Res 23: 1341–1349
6. Berger R, Florent G, Just M (1980) Decrease of the lymphoproliferative response to varicella-zoster virus antigen in the aged. Infect Immun 32: 24–27
7. Berger R, Trannoy E, Hollander G, Bailleux F, Rudin C, Creusvaux H (1998) A dose-response study of a live attenuated varicella-zoster virus (Oka strain) vaccine administered to adults 55 years of age and older. J Infect Dis 178 [Suppl] 1: S99–103
8. Cabirac GF, Mahalingam R, Wellish M, Gilden DH (1990) Trans-activation of viral tk promoters by proteins encoded by varicella zoster virus open reading frames 61 and 62. Virus Res 15: 57–68

9. Cohrs RJ, Barbour M, Gilden DH (1996) Varicella-zoster virus (VZV) transcription during latency in human ganglia: detection of transcripts mapping to genes 21, 29, 62, and 63 in a cDNA library enriched for VZV RNA. J Virol 70: 2789–2796

10. Cohrs RJ, Srock K, Barbour MB, Owens G, Mahalingam R, Devlin ME, Wellish M, Gilden DH (1994) Varicella-zoster virus (VZV) transcription during latency in human ganglia: construction of a cDNA library from latently infected human trigeminal ganglia and detection of a VZV transcript. J Virol 68: 7900–7908

11. Croen KD, Ostrove JM, Dragovic LJ, Straus SE (1988) Patterns of gene expression and sites of latency in human nerve ganglia are different for varicella-zoster and herpes simplex viruses. Proc Natl Acad Sci USA 85: 9773–9777

12. Debrus S, Sadzot-Delvaux C, Nikkels AF, Piette J, Rentier B (1995) Varicella-zoster virus gene 63 encodes an immediate-early protein that is abundantly expressed during latency. J Virol 69: 3240–3245

13. Defechereux P, Debrus S, Baudoux L, Rentier B, Piette J (1997) Varicella-zoster virus open reading frame 4 encodes an immediate-early protein with posttranscriptional regulatory properties. J Virol 71: 7073–7079

14. Defechereux P, Debrus S, Baudoux L, Schoonbroodt S, Merville MP, Rentier B, Piette J (1996) Intracellular distribution of the ORF4 gene product of varicella-zoster virus is influenced by the IE62 protein. J Gen Virol 77: 1505–1513

15. Defechereux P, Melen L, Baudoux L, Merville-Louis MP, Rentier B, Piette J (1993) Characterization of the regulatory functions of varicella-zoster virus open reading frame 4 gene product. J Virol 67: 4379–4385

16. Disney GH, McKee TA, Preston CM, Everett RD (1990) The product of varicella-zoster virus gene 62 autoregulates its own promoter. J Gen Virol 71: 2999–3003

17. Gilden DH, Rozenman Y, Murray R, Devlin M, Vafai A (1987) Detection of varicella-zoster virus nucleic acid in neurons of normal human thoracic ganglia. Ann Neurol 22: 377–380

18. Hayward AR, Herberger M (1987) Lymphocyte responses to varicella zoster virus in the elderly. J Clin Immunol 7: 174–178

19. Hope-Simpson RE (1965) The nature of herpes zoster: a long-term study and a new hypothesis. Proc R Soc Med 58: 9–20

20. Hyman RW, Ecker JR, Tenser RB (1983) Varicella-zoster virus RNA in human trigeminal ganglia. Lancet 2: 814–816

21. Inchauspe G, Nagpal S, Ostrove JM (1989) Mapping of two varicella-zoster virus-encoded genes that activate the expression of viral early and late genes. Virology 173: 700–709

22. Jugovic P, Hill AM, Tomazin R, Ploegh H, Johnson DC (1998) Inhibition of major histocompatibility complex class I antigen presentation in pig and primate cells by herpes simplex virus type 1 and 2 ICP47. J Virol 72: 5076–5084

23. Kinchington PR, Bookey D, Turse SE (1995) The transcriptional regulatory proteins encoded by varicella-zoster virus open reading frames (ORFs) 4 and 63, but not ORF 61, are associated with purified virus particles. J Virol 69: 4274–4282

24. Kinchington PR, Hougland JK, Arvin AM, Ruyechan WT, Hay J (1992) The varicella-zoster virus immediate-early protein IE62 is a major component of virus particles. J Virol 66: 359–366

25. Levin MJ, Barber D, Goldblatt E, Jones M, LaFleur B, Chan C, Stinson D, Zerbe GO, Hayward AR (1998) Use of a live attenuated varicella vaccine to boost varicella-specific immune responses in seropositive people 55 years of age and older: duration of booster effect. J Infect Dis 178 [Suppl 1]: S109–S112

26. Lungu O, Annunziato PW, Gershon A, Staugaitis SM, Josefson d, Larussa P, Silverstein SJ (1995) Reactivated and latent varicella-zoster virus in human dorsal root ganglia. Proc Natl Acad Sci USA 92: 10980–10984

27. Lungu O, Annunziato PW (1999) Varicella-zoster virus: latency and reactivation. In: Wolff MH, Schünemann S, Schmidt A (eds) Varicella-zoster virus. Molecular biology, pathogenesis and clinical aspects. Karger, Basel, pp 61–75

28. Lungu O, Panagiotidis CA, Annunziato PW, Gershon AA, Silverstein SJ (1998) Aberrant intracellular localization of varicella-zoster virus regulatory proteins during latency. Proc Natl Acad Sci USA 95: 7080–7085

29. Mahalingam R, Kennedy PGE, Gilden DH (1999) The problems of latent varicella eoster virus in human ganglia:precise cell location and viral content. J Neurovirol 5: 445–446

30. Mahalingam R, Wellish M, Cohrs R, Debrus S, Piette J, Rentier B, Gilden DH (1996) Expression of protein encoded by varicella-zoster virus open reading frame 63 in latently infected human ganglionic neurons. Proc Natl Acad Sci USA 93: 2122–2124

31. Meier JL, Holman RP, Croen KD, Smialek JE, Straus SE (1993) Varicella-zoster virus transcription in human trigeminal ganglia. Virology 193: 193–200

32. Moriuchi H, Moriuchi M, Straus SE, Cohen JI (1993) Varicella-zoster virus (VZV) open reading frame 61 protein transactivates VZV gene promoters and enhances the infectivity of VZV DNA. J Virol 67: 4290–4295

33. Nagpal S, Ostrove JM (1991) Characterization of a potent varicella-zoster virus-encoded trans-repressor. J Virol 65: 5289–5296

34. Nikkels AF, Debrus S, Sadzot-Delvaux C, Piette J, Rentier B, Pierard GE (1995) Immuno-histochemical identification of varicella-zoster virus gene 63-encoded protein (IE63) and late (gE) protein on smears and cutaneous biopsies: implications for diagnostic use. J Med Virol 47: 342–347

35. Patel Y, Gough G, Coffin RS, Thomas S, Cohen JI, Latchman DS (1998) Cell type specific repression of the varicella zoster virus immediate early gene 62 promoter by the cellular Oct-2 transcription factor. Biochim Biophys Acta 1397: 268–274

36. Pereira RA, Tscharke DC, Simmons A (1994) Upregulation of class I major histocompatibility complex gene expression in primary sensory neurons, satellite cells, and Schwann cells of mice in response to acute but not latent herpes simplex virus infection in vivo. J Exp Med 180: 841–850

37. Perera LP, Kaushal S, Kinchington PR, Mosca JD, Hayward GS, Straus SE (1994) Varicella-zoster virus open reading frame 4 encodes a transcriptional activator that is functionally distinct from that of herpes simplex virus homology ICP27. J Virol 68: 2468–2477

38. Perera LP, Mosca JD, Sadeghi-Zadeh M, Ruyechan WT, Hay J (1992) The varicella-zoster virus immediate early protein, IE62, can positively regulate its cognate promoter. Virology 191: 346–354

39. Sadzot-Delvaux C, Debrus S, Nikkels A, Piette J, Rentier B (1995) Varicella-zoster virus latency in the adult rat is a useful model for human latent infection. Neurology 45: S18–S20

40. Sadzot-Delvaux C, Kinchington PR, Debrus S, Rentier B, Arvin AM (1997) Recognition of the latency-associated immediate early protein IE63 of varicella-zoster virus by human memory T lymphocytes. J Immunol 159: 2802–2806

41. Shiraki K, Hyman RW (1987) The immediate early proteins of varicella-zoster virus. Virology 156: 423–426

42. Tomazin R, Hill AB, Jugovic P, York I, van Endert P, Ploegh HL, Andrews DW, Johnson DC (1996) Stable binding of the herpes simplex virus ICP47 protein to the peptide binding site of TAP. EMBO J 15: 3256–3266

43. Vafai A, Murray RS, Wellish M, Devlin M, Gilden DH (1988) Expression of varicella-zoster virus and herpes simplex virus in normal human trigeminal ganglia. Proc Natl Acad Sci USA 85: 2362–2366
44. Whitley RJ (1996) Herpes simplex virus. In: Fields BN, Knipe DM, Howley PM (eds) Virology. Lippincott-Raven, New York, pp 2297–2342
45. Wroblewska Z, Valyi-Nagy T, Otte J, Dillner A, Jackson A, Sole DP, Fraser NW (1993) A mouse model for varicella-zoster virus latency. Microb Pathog 15: 141–151

Authors' address: C. Sadzot-Delvaux, Department of Microbiology, Fundamental Virology, Pathology B23, Liège University, B-4000 Sart Tilman-Liège, Belgium.

Mutagenesis of the varicella-zoster virus genome: lessons learned

J. I. Cohen

Medical Virology Section, Laboratory of Clinical Investigation,
National Institutes of Health, Bethesda, Maryland, U.S.A.

Summary. The varicella-zoster virus (VZV) genome encodes at least 70 genes. We have developed a cosmid based system to inactivate individual viral genes or to insert foreign genes into the genome. We have shown that many VZV genes are not required for replication of the virus in cell culture. Several of these genes, including VZV ORF61, ORF47, and ORF10, have unexpected phenotypes in cell culture and differ from their homologs in the better studied herpes simplex virus (HSV). We have also used the Oka strain of VZV as a live virus vaccine vector. Guinea pigs vaccinated with recombinant VZV expressing HSV-2 glycoprotein D and challenged with HSV-2 have reduced severity of primary genital herpes and reduced mortality compared to animals receiving parental VZV. Recently we have inserted the human immunodeficiency virus (HIV) and simian immunodeficiency virus (SIV) glycoprotein 160 genes into the Oka strain of VZV and have shown that these proteins are expressed in recombinant virus-infected cells. Thus, directed mutagenesis of the VZV genome is providing new insights into viral pathogenesis and may provide new candidate vaccines.

Introduction

While the genome of varicella-zoster virus (VZV), only 125 kilobase pairs of DNA, is the smallest of the human herpesviruses, genetic studies of the virus are less extensive than the better studied and genetically more complex herpes simplex virus (HSV), human cytomegalovirus, and Epstein-Barr virus. The major reason for the lag in studies of VZV is the difficulty in obtaining cell-free virus for plaque purification of viral mutants from wild-type virus. About seven years ago we developed a system to engineer specific mutations into the VZV genome by transfection of cosmid DNAs.

Four cosmids were obtained from the Oka VZV vaccine virus that span the entire genome. Transfection of melanoma cells with these cosmids results in homologous recombination of the cosmids with production of infectious VZV. The replication properties of the recombinant virus resemble that of the parental

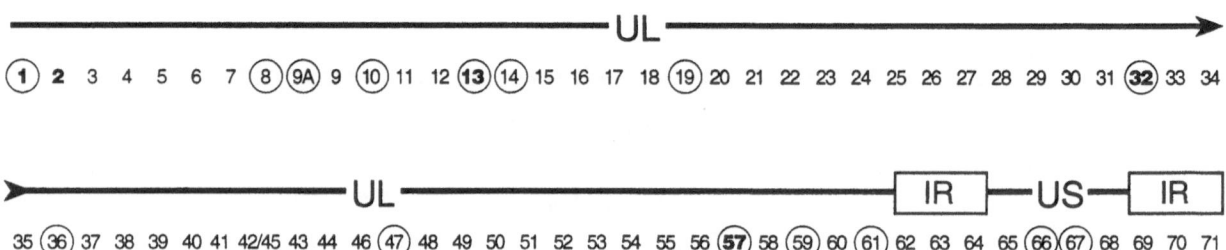

Fig. 1. VZV genes dispensable for virus replication in vitro. Of the 69 unique genes in the virus, 15 have been shown not to be required for VZV replication in vitro (circled). Genes that do not have homologs in HSV are bolded

virus from which it was obtained [2]. Transfection of cells with cosmids containing deletions or stop codons within specific viral genes results in production of viruses with these specific mutations. Alternatively, insertion of a cassette containing a foreign DNA with a eukaryotic promoter results in expression of the foreign gene in virus-infected cells.

Two groups have now reported the generation of VZV mutants using the cosmid system [2, 9]. Mutagenesis studies have shown that 15 different viral genes are dispensable for VZV replication in cell culture (circled in Fig. 1). Most of the targeted genes have homologs in the better studied HSV. However, some of these genes (bold in Fig. 1) are unique to VZV. Nearly all the genes tested, including those without homologs in HSV (open reading frame [ORF] 1, 13, 32, 57), have been shown to be dispensable for VZV replication. However, one VZV gene (ORF5, gK) was proven to be essential for replication [10]. Inactivation of three VZV genes (ORF9A [21], ORF61 [1], or ORF67 [9]) results in impaired syncytial formation in vitro. Inactivation of the VZV protein kinases, that are dispensable for replication in fibroblasts, result in a marked impairment for replication in human fetal T cells in the SCID-hu mouse [12]. This review will focus on three of the VZV mutants that have been constructed in our laboratory and the unexpected phenotypes that have been observed. In addition, the use of VZV as a vaccine vector for expression of foreign DNAs will be discussed.

Varicella-zoster virus ORF61

VZV ORF61 is a putative immediate-early gene that encodes a phosphoprotein [25]. ORF61 protein contains a RING finger domain near its amino terminus [4]. This domain consists of cysteine and histidine residues in a specific pattern, that is highly conserved with other alphaherpesviruses. The RING finger of ORF61 is known to bind zinc and the RING finger of HSV ICP0, the homolog of ORF61, is important for interactions with other proteins. ORF61 protein transactivates VZV immediate-early and early promoters and enhances the infectivity of viral DNA [17]. While full length ORF61 protein acts as a transactivator, successive carboxy truncation mutants result in loss of transactivation, and further truncations act as transrepressors [13]. Each of the transrepressing mutants contains the RING finger domain; when the RING finger is deleted, there is no transrepression.

VZV with a large deletion in ORF61, which includes loss of the RING finger domain, can be propagated in cell culture. Since carboxy terminal mutants of ORF61 containing the RING finger domain act as transrepressors, a virus expressing a truncated ORF61 protein might inhibit the ability of the virus to replicate. Multiple attempts to produce VZV with a truncation in ORF61 protein after the RING finger domain were unsuccessful [1]. In contrast, an HSV mutant with a similar truncation after the RING finger of ICP0, the HSV homolog of ORF61, could be propagated in cell culture and replicated at levels similar to an ICP0 deletion mutant [24]. Thus, VZV ORF61 and HSV ICP0 truncation mutants have different effects on virus growth in vitro.

Analysis of the growth of the VZV ORF61 large deletion mutant showed that it is partially impaired for growth in melanoma cells, but severely impaired for growth in schwannoma cells [1]. Cells infected with the VZV ORF61 deletion mutant have abnormal syncytia formation in vitro and express reduced levels of VZV glycoprotein E.

Varicella-zoster virus ORF47

VZV ORF47 encodes a protein kinase that phosphorylates itself and the ORF62 protein in vitro [19]. ORF47 protein can also phosphorylate casein in vitro [18]. ORF47 protein is located in the capsid and/or tegument of the virion [26].

Cells infected with a mutant VZV that is unable to express ORF47 protein are still able to phosphorylate ORF62 protein to a similar extent as cells infected with parental virus [5]. However, the phosphorylation of several proteins is altered in cells lacking the ORF47 protein. One of these proteins, encoded by VZV ORF32, shows reduced phosphorylation in cells infected with VZV unable to express ORF47 when compared with parental virus [20].

VZV unable to express ORF47 protein replicates to titers similar to those of parental virus in human melanoma cells and fibroblasts. However, the ORF47 protein is essential for VZV replication in human fetal T cell and skin implants in SCID mice [12]. Recent studies have confirmed that ORF47 is required for VZV replication in human umbilical cord blood T cells in vitro (23a).

VZV ORF47 is homologous to the UL13 protein kinase of HSV [23]. While ORF47 is not required for the phosphorylation of ORF63 [5], HSV UL13 phosphorylates ICP22, the HSV homolog of VZV ORF63. In addition, while VZV replicates in human fetal T cells in the SCID mouse, HSV replicates in fetal epithelial but not T cells [11]. Thus, while VZV and HSV both encode homologous protein kinases, and ORF47 is essential for replication of the virus in T cells, the homologous HSV protein (UL13 protein kinase) is insufficient to allow HSV to replicate in fetal T cells in this model.

Varicella-zoster virus ORF10

VZV ORF10 is a putative late gene that encodes a potent viral transactivator. ORF10 protein upregulates the expression of the VZV immediate-early ORF62 protein and enhances the infectivity of VZV DNA [16]. The transactivation domain of ORF10 is located near the amino terminus of the protein [15]. The ORF10 protein is located in the viral tegument [7]. VZV ORF10 is the homolog of HSV VP16, also referred to as alpha trans-inducing factor or Vmw65. Both proteins are potent transactivators and form complexes [14, 27] with the cellular proteins Oct1 and HCF to bind at TAATGARAT or TAATGARAT-like sequences located upstream of the major immediate-early protein of their viruses (ORF62 protein for VZV; ICP4 for HSV).

VZV with stop codons inserted into ORF10 or deleted for ORF10 is able to replicate to titers identical to those seen with parental virus in vitro [3]. VZV virions deleted for ORF10 are indistinguishable from the parental virus by electron microscopy. In marked contrast, HSV deleted for VP16 is unable to replicate in cell culture unless complemented by a cell line expressing the protein [28]. Because the HSV mutant has a defect in capsid maturation and assembly, the finding that ORF10 is dispensable for replication in vitro indicates that these two proteins have different structural functions.

VZV as a vaccine vector

Recent studies have examined the ability of the attenuated Oka vaccine virus to serve as a live virus vector to deliver foreign proteins to the immune system. VZV has a number of potential advantages in this regard. The Oka VZV vaccine is approved for vaccination of children in the United States and is safe except in severely immunocompromised persons. Infection with VZV induces both VZV-specific antibodies and MHC class I and II restricted cytotoxic T cells. The VZV genome can accommodate large inserts of foreign DNA. On the other hand, VZV has some disadvantages as a live vaccine vector. Most adolescents and adults are immune to VZV and therefore the vaccine vector would have to be given during childhood. Few doses of virus can be given, because the immune response to VZV would limit the amount of virus replication with subsequent vaccinations. Finally, because the virus becomes latent in neuronal cells, it is uncertain whether reactivation in these cells with expression of an immunogenic foreign protein would create a problem.

Previous studies showed that two foreign genes, the Epstein-Barr virus glycoprotein gp350 [8] and the hepatitis B surface antigen [22], could be inserted into the VZV genome and expressed in virus-infected cells. However, neither of these studies showed that the recombinant virus could protect against challenge with the foreign virus.

We inserted the HSV-2 glycoprotein D (gD) gene into the VZV genome using the cosmid system [6]. HSV-2 gD is a major target for neutralizing antibody and T cell responses to HSV-2. Cells infected with recombinant VZV expressing HSV-2 gD expressed gD on their surface and on the surface of virions. Guinea pigs were immunized with either cells containing the Oka virus expressing HSV-2 gD or the parental virus. Animals that received VZV expressing HSV-2 gD developed neutralizing antibody titers to HSV-2. Although all of the animals became infected after challenge with HSV-2, the severity of disease was diminished in animals receiving VZV expressing HSV-2 gD compared with those receiving parental VZV. The absence of protection from infection was thought to be due to the relatively low levels of gD expressed by the recombinant virus or to the low level of VZV replication in guinea pigs.

Recently we constructed VZV genomes containing either the HIV gp160 or SIV gp160 genes. Cells infected with these viruses express proteins of 120 and 160 kDa. The retroviral proteins are expressed in the cytoplasm and on the membranes of recombinant VZV-infected cells. In collaboration with Dr. Mark Feinberg at Emory University, we inoculated rhesus monkeys with the viruses and currently are analyzing the immune responses to the foreign proteins.

The ability to readily mutate or delete individual genes from the VZV genome is providing important new information on the functions of viral proteins in the context of the virus. Furthermore, the ability to insert new genes into the viral genome may result in novel vaccines that might be effective for protection against both VZV and additional pathogens.

References

1. Cohen JI, Nguyen H (1998) Varicella-zoster ORF61 deletion mutants replicate in cell culture, but a mutant with stop codons in ORF61 reverts to wild-type virus. Virology 246: 306–316
2. Cohen JI, Seidel KE (1993) Generation of varicella-zoster virus (VZV) and viral mutants from cosmid DNAs: VZV thymidylate synthetase is not essential for replication in vitro. Proc Natl Acad Sci USA 90: 7376–7380
3. Cohen JI, Seidel KE (1994) Varicella-zoster virus (VZV) open reading frame 10 protein, the homolog of the essential herpes simplex virus protein VP16, is dispensable for VZV replication in vitro. J Virol 68: 7850–7858
4. Everett RD, Barlow P, Milner A, Luisi B, Orr A, Hope G, Lyon DA (1993) A novel arrangement of zinc-binding residues and secondary structure in the C_3HC_4 motif of an alpha herpes virus protein family. J Mol Biol 234: 1038–1047
5. Heineman TC, Cohen JI (1995) The varicella-zoster virus (VZV) open reading frame 47 (ORF47) protein kinase is dispensable for viral replication and is not required for phosphorylation of ORF63 protein, the VZV homolog of herpes simplex virus ICP22. J Virol 69: 7367–7370
6. Heineman TC, Connelly BL, Bourne N, Stanberry LR, Cohen JI (1995) Immunization with recombinant varicella-zoster virus expressing herpes simplex virus type 2 glycoprotein D reduces the severity of genital herpes in guinea pigs. J Virol 69: 8109–8113
7. Kinchington PR, Hougland JK, Arvin AM, Ruyechan WT, Hay J (1992) The varicella-zoster virus immediate-early protein IE62 is a major component of virus particles. J Virol 66: 359–366
8. Lowe RS, Keller PM, Keech BJ, Davison AJ, Whang Y, Morgan AJ, Kieff E, Ellis RW (1987) Varicella-zoster virus as a live vector for the expression of foreign genes. Proc Natl Acad Sci USA 84: 3896–3900
9. Mallory S, Sommer M, Arvin AM (1997) Mutational analysis of the role of glycoprotein I in varicella-zoster virus replication and its effects on glycoprotein E conformation and trafficking. J Virol 71: 8279–8288
10. Mo C, Suen J, Sommer M, Arvin A (1999) Characterization of varicella-zoster virus glycoprotein K (open reading frame 5) and its role in virus growth. J Virol 73: 4197–4207

11. Moffat JF, Zerboni L, Kinchington PR, Grose C, Kaneshima H, Arvin AM (1998) Attenuation of the vaccine Oka strain of varicella-zoster virus and role of glycoprotein C in alphaherpesvirus virulence demonstrated in the SCID-hu mouse. J Virol 72: 965–974

12. Moffat JF, Zerboni L, Sommer MH, Heineman TC; Cohen JI, Kaneshima H, Arvin AM (1998) The ORF47 and ORF66 putative protein kinases of varicella-zoster virus determine tropism for human T cells and skin in the SCID-hu mouse. Proc Natl Acad Sci USA 95: 11969–11974

13. Moriuchi H, Moriuchi M, Cohen JI (1994) The RING finger domain of the varicella-zoster virus open reading frame 61 protein is required for its transregulatory functions. Virology 205: 238–246

14. Moriuchi H, Moriuchi M, Cohen JI (1995) Proteins and cis-acting elements associated with transactivation of the varicella-zoster virus (VZV) immediate-early gene 62 promoter by VZV open reading frame 10 protein. J Virol 69: 4693–4701

15. Moriuchi H, Moriuchi M, Pichyangkura R, Triezenberg SJ, Straus SE, Cohen JI (1995) Hydrophobic cluster analysis predicts an amino-terminal domain of varicella-zoster virus open reading frame 10 required for transcriptional activation. Proc Natl Acad Sci USA 92: 9333–9337

16. Moriuchi H, Moriuchi M, Straus SE, Cohen JI (1993) Varicella-zoster virus open reading frame 10 protein, the herpes simplex virus VP16 homolog, transactivates herpesvirus immediate-early gene promoters. J Virol 67: 2739–2746

17. Moriuchi H, Moriuchi M, Straus SE, Cohen JI (1993) Varicella-zoster virus (VZV) open reading frame 61 protein transactivates VZV gene promoters and enhances the infectivity of VZV DNA. J Virol 67: 4290–4295

18. Ng TI, Grose C (1992) Serine protein kinase associated with varicella-zoster virus ORF 47. Virology 191: 9–18

19. Ng TI, Keenan L, Kinchington PR, Grose C (1994) Phosphorylation of varicella-zoster virus open reading frame (ORF) 62 regulatory product by viral ORF47-associated protein kinase. J Virol 68: 1350–1359

20. Reddy SM, Cox E, Iofin I, Soong W, Cohen JI (1998) Varicella-zoster virus (VZV) ORF32 encodes a phosphoprotein that is posttranslationally modified by the VZV ORF47 protein kinase. J Virol 72: 8083–8088

21. Ross J, Williams M, Cohen JI (1997) Disruption of the varicella-zoster virus dUTPase and the adjacent ORF9A gene results in impaired growth and reduced syncytia formation in vitro. Virology 234: 186–195

22. Shiraki K, Hayakawa Y, Mori H, Namazue J, Takamizawa A, Yoshida I, Yamanishi K, Takahashi M (1991) Development of immunogenic recombinant Oka varicella vaccine expressing hepatitis B surface antigen. J Gen Virol 72: 1393–1399

23. Smith RF, Smith TF (1989) Identification of new protein kinase-related genes in three herpesviruses, herpes simplex virus, varicella-zoster virus, and Epstein-Barr virus. J Virol 63: 450–455

23a. Soong W, Shultz JC, Patera AC, Sommer MH, Cohen JI (2000) Infection of human T lymphocytes with varicella-zoster virus: an analysis with viral mutants and clinical isolates. J Virol 74: 1864–1870

24. Spatz SJ, Nordby EC, Weber PC (1997) Construction and characterization of a recombinant herpes simplex virus type 1 which overexpresses the transrepressor protein ICP0R. Virology 228: 218–228

25. Stevenson D, Colman KL, Davison AJ (1992) Characterization of the varicella-zoster virus gene 61 protein. J Gen Virol 73: 521–530

26. Stevenson D, Colman KL, Davison AJ (1994) Characterization of the putative protein kinases specified by varicella-zoster virus genes 47 and 66. J Gen Virol 75: 317–326
27. Xiao P, Capone JP (1990) A cellular factor binds to the herpes simplex virus type 1 transactivator Vmw65 and is required for Vmw65-dependent protein-DNA complex assembly with Oct-1. Mol Cell Biol 10: 4974–4977
28. Weinheimer SP, Boyd BA, Durham SK, Resnick JL, O'Boyle DR (1992) Deletion of the VP16 open reading frame of herpes simplex virus type 1. J Virol 66: 258–269

Authors' address: Dr. J. I. Cohen, Medical Virology Section, Laboratory of Clinical Investigation, National Institutes of Health, Bethesda, MD 20892, U.S.A.

Immune evasion mechanisms of varicella-zoster virus

A. Abendroth* and **A. Arvin**

Stanford University School of Medicine, Stanford, California, U.S.A.

Summary. Varicella-zoster virus can to modulate the expression of class I and class II major histocompatibility (MHC) molecules. MHC class I expression is downregulated in VZV-infected T cells as well as in fibroblasts. VZV-infected cells do not respond to exposure to interferon-γ (IFN-γ) by upregulation of MHC class II expression. However, MHC class II expression is induced when cells are treated with IFN-γ before VZV infection. These effects on MHC class I and class II expression can be expected to interfere transiently with adaptive immune responses of the host, mediated by CD4 and CD8 T cells, ensuring that the virus has sufficient opportunity for transmission to susceptible contracts.

Introduction

Varicella zoster virus (VZV) is a human herpesvirus that causes varicella (chickenpox) as the primary infection in susceptible individuals, establishes latency in sensory ganglia and may reactivate as herpes zoster (shingles) [5, 9]. Both innate cell-mediated responses, and antigen-specific immunity are elicited during the course of primary VZV infection. The early host responses to VZV are nonspecific and involve natural killer (NK) cells and interferons (IFN) that function to restrict virus replication and spread [6]. VZV specific T cell recognition is critical for host recovery from varicella and both major histocompatibility complex (MHC) class I restricted CD8[+] and MHC class II restricted CD4[+] T cells are sensitized during primary VZV infection. VZV specific CD4[+] T cells that are induced during primary infection are predominantly of the Th1 type [42] and function to produce high levels of IFN-γ, which potentiates the clonal expansion of VZV specific T cells [5]. Although the classical cytotoxic T lymphocyte (CTL) response is mediated by CD8[+] T cells that recognize viral peptides in association with MHC class I molecules, VZV-specific CTLs can also exhibit MHC class II (CD4[+]) restricted killing of infected target cells [14, 15, 22, 38]. Based upon

---

*Current address: Centre for Virus Research, Westmead Millennium, Institute of Health Research, University of Sydney, Westmead, NSW, Australia.

these observations, immunomodulatory mechanisms that limit the initial presentation of VZV peptides by MHC class I or class II pathways are likely to have an important effect on viral pathogenesis.

Immune evasion mechanisms of VZV resulting in downregulation of MHC class I molecules

The MHC class I complex is a heterodimer, with a membrane bound heavy chain (αC) and a light chain $\beta_2$microglobulin (β_2m). MHC class I complexes present peptides derived from cytosolic proteins to CD8 T cells. In this pathway, antigenic peptides generated by cytosolic proteases are transported into the endoplasmic reticulum (ER) by the ATP-dependent transporter associated with antigen processing (TAP), and associate with MHC I heterodimers. The peptide-MHC complex is transported from the ER through the Golgi compartment to the cell membrane. When expressed with MHC class I, the peptide is recognized by cytotoxic CD8 T cells.

Our investigations have revealed that VZV downregulates cell-surface MHC I expression in cultured human cells [1]. Cells were stained 24 h after infection using polyclonal VZV-immune serum and mouse monoclonal antibody to MHC I antigens, and analysed by flow cytometry. Among VZV negative cells, 90% were MHC I positive, whereas only 25% of the VZV positive cells expressed MHC I. Control experiments using staining for transferrin receptor showed that more than 98% of both VZV positive and VZV negative cells had this cell surface protein. Similar effects were observed using primary human fetal lung fibroblast (MRC-5) cells and transformed human melanoma cells. Experiments using thymus/liver implants in SCIDhu mice inoculated with VZV showed that the virus decreased MHC class I expression on VZV-infected T cells from all of five mice that were examined. Diminished expression of MHC class I was observed on all VZV-infected T cells from CD8$^+$, CD4$^+$ and immature CD4$^+$CD8$^+$ T cell subpopulations.

When the intracellular localization of MHC I molecules was examined by immunofluorescence and confocal microscopy, we found that VZV-infected cells had MHC class I protein accumulation in the perinuclear region. Uninfected cells showed normal distribution of MHC class I molecules. Using pulse chase and immunoprecipitation in the presence of endo H, we found that MHC class I molecules acquired endo H resistance at the 3 h 'chase', whether cells were VZV infected or mock infected.

When intracellular probes for ER were used along with MHC I and VZV specific antibodies and examined by confocal microscopy, the dual colour image did not reveal any colocalization in VZV infected cells. However, MHC class I colocalized with ceramide staining for Golgi in VZV infected cells. To identify the phase of viral gene expression associated with MHC I downregulation, we added phosphonacetic acid (PAA) an inhibitor of viral DNA replication and thus late gene expression, to VZV-infected cells and determined cell-surface MHC class I expression by flow cytometry. Although expression of gC was blocked by

PAA, MHC class I was downregulated with or without PAA treatment, suggesting that a VZV immediate early or early gene product(s), or a virion component(s) acts to inhibit transport of MHC class I protein to cell surfaces.

Immune evasion mechanisms of VZV resulting in interference with IFN-γ induced upregulation of MHC class II molecules

MHC class II proteins are polymorphic heterodimers with a and b chains. The MHC class II molecules function to present exogenous derived peptides to CD4$^+$ T cells. In the case of MHC class II proteins, the a and b chains form a heterodimer in the ER. This complex associates with the invariant chain (Ii) and is transported through the Golgi and *trans*-Golgi reticulum to endosomes in the cytoplasm. The Ii chain is degraded in these endosomes, allowing antigenic peptides, which are produced by proteolysis of endocytosed proteins, to bind to the a/b heterodimer. The peptide-MHC class II complex then appears on the cell surface [32]. In contrast to MHC class I, MHC class II proteins are expressed constitutively only on B cells, monocytes, dendritic cells and thymic epithelium. IFN-γ serves an important function in allowing CD4 T cell recognition of infected cells because it is a potent inducer of MHC class II expression on many cell types. Many viruses including adenoviruses, murine and human cytomegaloviruses, mouse hepatitis virus, human immunodeficiency virus (HIV), Kirsten murine sarcoma virus and measles virus have mechanisms to inhibit IFN-γ induced upregulation of MHC class II expression [8, 16, 17, 27–29, 33, 36, 37]. In most cases, viral proteins

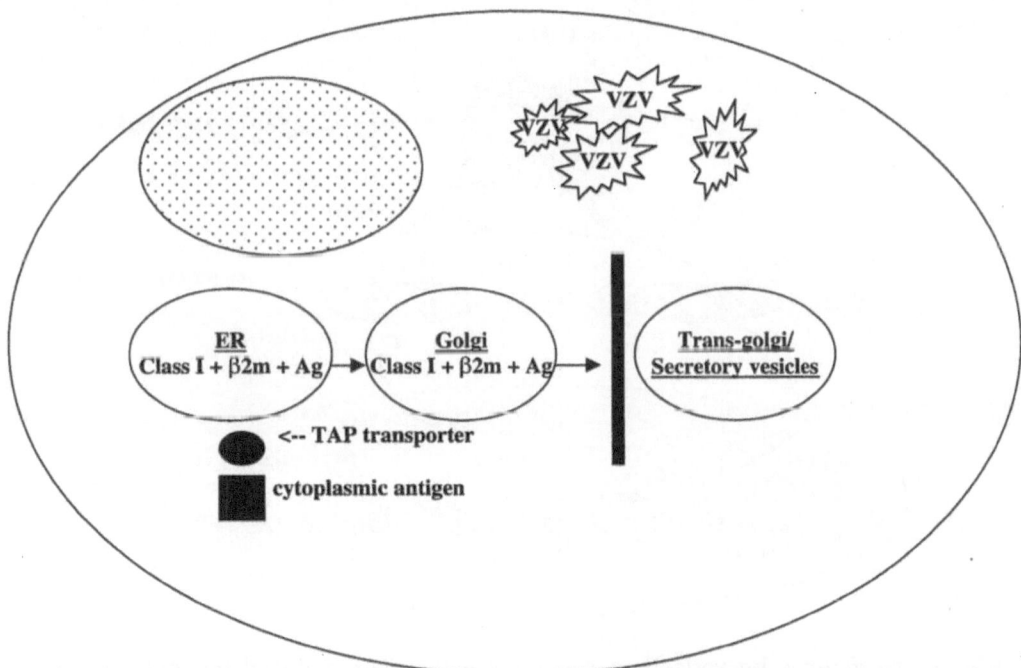

Fig. 1. Interference by varicella-zoster virus MHC class I expression

block MHC class I expression at the level of mRNA transcription, but may also modulate via post-transcriptional effects.

Our experiments used fibroblasts and flow cytometry analysis to investigate the effect of VZV on IFN-γ stimulated MHC class II expression in fibroblasts. Twelve hours after infection, fibroblasts were treated with human IFN-γ or 36 h and stained for VZV and MHC class II expression using polyclonal VZV-immune serum and mouse monoclonal antibody to MHC class II DR-a. Only 26% of VZV infected cells expressed MHC class II compared with 86% of the uninfected cells; transferrin receptor expression was 98% in both VZV infected and uninfected cells. In contrast, when fibroblasts were treated with IFN-γ for 36 h before VZV infection and analyzed at 46 h, 85% of VZV infected and uninfected cells expressed MHC class II.

Because IFN-γ induction of MHC class II expression occurs at the level of gene transcription [7], we investigated IFN-γ induced upregulation of MHC class II RNA expression by northern blot and in situ hybridization. MHC class II DR-a transcripts were detected uninfected cells treated with IFN-γ whereas VZV-infected cells had no detectable MHC class II DR-a transcripts. Cells infected with VZV and treated with IFN-γ were analysed by in situ hybridization with probes for MHC class II DR-γ transcripts. MHC class II DR-γ transcripts were detected in most uninfected cells but not in VZV infected cells that were MHC class II negative. MHC class II transcripts could not be found in VZV infected cells that lacked MHC class II expression on cell surfaces.

In addition to these experiments using tissue culture cells, we assessed the distribution of MHC class II positive cells in skin biopsies of VZV lesions. VZV

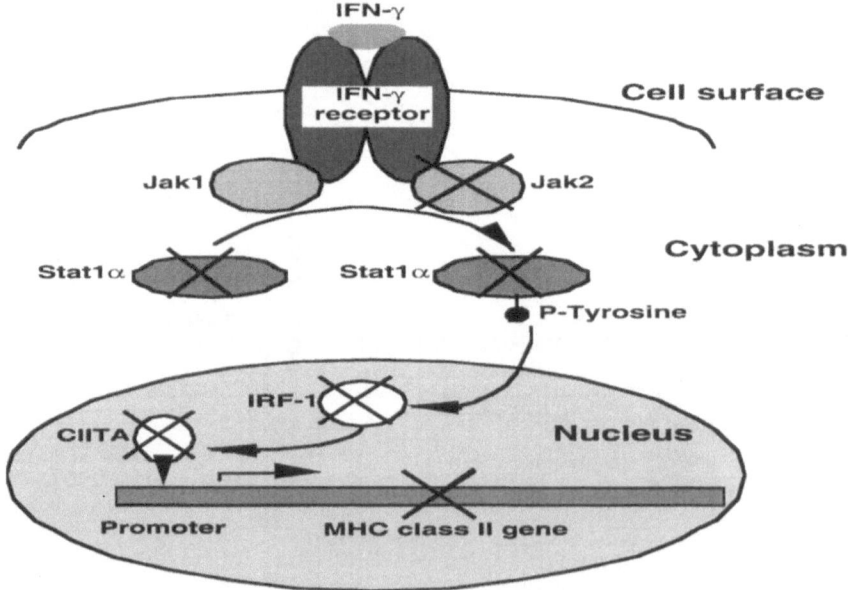

Fig. 2. Interference by varicella-zoster virus with IFN-γ induced upregulation of MHC class II expression via the Stat signal transduction pathway. Reprinted with permission from [1]

IE62 transcripts were detectable within cells in areas of the lesions but MHC class II DR-α transcripts were detected only in infiltrating inflammatory cells. These experiments indicate that MHC class II is not expressed in VZV-infected cells in vivo, but that transcripts can be found in cells near infected cells during early cutaneous lesion formation.

Interferon regulatory factor-1 (IRF-1) is required to transactivate MHC class II induced by IFN-γ. IRF-1 RNA was detected in IFN-γ treated uninfected cells but not those that were VZV-infected and which lacked MHC class II DR-α. Stat 1α induces IRF-1 expression in response to IFN-γ. This protein is a component of the Jak/Stat signal transduction pathway, which includes Jak1 and Jak2. To determine whether the Jak/Stat pathway was disrupted in VZV-infected cells, we used western blot to assess Jak1, Jak2 and Stat 1α protein expression. Levels of Jak1 protein expression did not change but Jak2 and Stat1α protein expression were reduced significantly in cells that were VZV positive and negative for MHC class II expression after IFN-γ treatment. These observations indicate that VZV interferes with the Jak/Stat signal transduction pathway by inhibiting Jak2 and Stat1α.

Discussion

Taken together, these experiments demonstrate that VZV encodes genes that can facilitate immune evasion, whether the host response is mediated by CD5[+] or CD8[+] T cells. In the case of MHC class I, this interference allows VZV-infected cells to avoid immediate clearance by CD8 T cells. The finding that MHC I molecules accumulated in the Golgi complex of VZV-infected cells indicates that the mechanism of VZV immunomodulation of MHC class I expression differs from those of HSV and human cytomegalovirus (HCMV) [20]. In HSV-infected cells, MHC class I molecules are retained in the endoplasmic reticulum by inter- action of ICP47 with the TAP complex, blocking TAP-mediated antigen transport [2, 11, 13, 21, 39, 41]. HCMV has several proteins that alter MHC class I assembly by preventing TAP-mediated antigen transport, retaining MHC class I molecules in the ER and facilitating cytosolic degradation of the MHC class I complex [2, 3, 18, 19, 23–25, 30, 40]. MCMV inhibits transport of MHC class I molecules past the cis-Golgi [43]. Thus VZV differs from HSV and HCMV in its methods of reducing the surface expression of MHC class I proteins. The viral protein or proteins that modulate MHC class I expression have not been identified but it must be unique because VZV has no homologues for HSV ICP47 or for pro- teins from other herpesviruses that interfere with MHC class I expression. Based on the retention of MHC class I molecules in the Golgi in VZV-infected cells, some viralprotein (or proteins) probably binds to a component of the MHC class I complex [2, 11, 13, 21, 24–26, 30, 35, 39–41].

CD4 T cells that recognize VZV are almost exclusively of the Th1 type, and IFN-γ is a major cytokine product of these memory T cells. The ability of VZV to inhibit MHC class II expression in most infected cells, despite exposure to high concentrations of IFN-γ, provides a mechanism by which the virus can limit antiviral activity of CD4$^+$ T cells. Any viral immunomodulatory effect that slows the initial clonal amplification of antigen-specific CD4 T cells may be expected to facilitate VZV replication at cutaneous sites. VZV specific CD4 T cells also mediate MHC class II restricted lysis of cells that express VZV proteins [6]. This cytotoxic capacity of CD4 T cells should also be inhibited, unless MHC class II upregulation can be triggered by IFN-γ. The analysis of skin biopsies showed that dermal and epidermal cells infected with VZV did not express MHC class II transcripts at early stages of lesion formatin. Since uninfected cells remained susceptible to IFN-γ induced upregulation of MHC class II, IFN-γ may allow CD4 T cells to eliminate cells that become infected as the inflammatory response progresses. VZV inhibition of IFN-γ dependent transcription of the MHC class I DRα gene resembles observations about the inhibition of MHC class II expression by HCMV. However, HCMV inhibits MHC class II expression in human fibroblasts at the level of Jak/Stat signal transduction, by specifically decreasing Jak1 expression while VZV reduced Jak2 expression.

Conclusion

Both CD4 and CD8 T cell responses are acquired during primary VZV infection, and antigen-specific memory T cells persist during latency and increase during the resolution of recurrent VZV infections [6]. It is therefore useful for the virus to modulate expression of MHC class I and class II proteins to avoid immune surveillance. While the effects are incomplete, blocking the effector arm of the host response is likely to facilitate VZV replication in skin lesions. Whether the infection is primary or recurrent, cutaneous lesions are a resevoir for spread to susceptible individuals. From the perspective of achieving viral persistence in the population over time, the immunomodulatory effects of VZV function optimally if they are transient and limited, so that the host recovers and latency, with the potential for later reactivation, is established. VZV appears to have evolved gene products that can achieve this outcome, which has allowed the virus to persist within the host population for millions of years.

Acknowledgement

This work was supported by NIH grant AI20459.

References

1. Abendroth A, Slobedman B, Lee E, Wallace M, Mellins E, Arvin A (2000) Modulation of major histocompatibility complex class II expression by varicella zoster virus. J Virol 74: 1900–1907
2. Ahn K, Angulo A, Ghazal P, Peterson PA, Yang Y, Fruh K (1996) Human cytomegalovirus inhibits antigen presentation by a sequential multistep process. Proc Natl Acad Sci USA 93: 10990–10995
3. Ahn K, Gruhler A, Galocha B, Jones TR, Wiertz EJ, Ploegh HL, Peterson PA, Yang Y, Fruh K (1997) The ER-luminal domain of the HCMV glycoprotein US6 inhibits peptide translocation by TAP. Immunity 6: 613–621

4. Ahn K, Meyer TH, Uebel S, Sempe P, Djaballah H. Yang Y, Peterson PA, Fruh K, Tampe R (1996) Molecular mechanism and species specificity of TAP inhibition by herpes simplex virus ICP47. EMBO J 15: 3247–3255

5. Arvin A (1995) Varicella-zoster virus. In: Fields B, Knipe D, Howley P (eds) Fields Virology. Raven, New York, pp 2547–2586

6. Arvin AM (1996) Immune responses to varicella-zoster virus. Infect Dis Clin North Am 10: 529

7. Boss JM (1997) Regulation of transcription of MHC class II genes. Curr Opin Immunol 9: 107–113

8. Buchmeier MA, Cooper NR (1989) Suppression of monocyte functions by human cytomegalovirus. Immunology 66: 278–283

9. Cohen J, Straus S (1995) Varicella zoster virus and its replication. In: Fields B, Knipe D, Howley P (eds) Fields Virology. Raven, New York, pp 2525–2546

10. Cohen JI (1998) Infection of cells with varicella-zoster virus down-regulates surface expression of class I major histocompatibility complex antigens. J Infect Dis 177: 1390–1393

11. Fruh K, Ahn K, Djaballah H, Sempe P, van Endert PM, Tampe R, Peterson PA, Yang Y (1995) A viral inhibitor of peptide transporters for antigen presentation. Nature 375: 415–418

12. Fruh K, Gruhler A, Krishna R, Schoenhals G (1999) A comparison of viral immune escape strategies targeting the MHC class I assembly pathway. Immunol Rev 168: 157–166

13. Galocha B, Hill A, Barnett B, Dolan A, Raimondi A, Cook R, Brunner J, McGeoch D, Ploegh H (1997) The active site of ICP47, a herpes simplex virus-encoded inhibitor of major histocompatibility complex (MHC)-encoded peptide transport associated with antigen processing (TAP), maps to the NH2-terminal 35 residues. J Exp Med 185: 1565–1572

14. Hayward A, Giller R, Levin M (1989) Phenotype, cytotoxic, and helper functions of T cells from varicella zoster virus stimulated cultures of human lymphocytes. Viral Immunol 2: 175–184

15. Hayward AR, Pontesilli O, Herberger M, Laszlo M, Levin M (1986) Specific lysis of varicella zoster virus-infected B lymphoblasts by human T cells. J Virol 58: 179–184

16. Heise MT, Connick M, Virgin HW (1998) Murine cytomegalovirus inhibits interferon gamma-induced antigen presentation of CD4 T cells by macrophages via regulation of expression of major histocompatibility comlex class II-associated genes. J Exp Med 187: 1037–1046

17. Heise MT, Pollock JL, O'Guin A, Barkon ML, Bormley S, Virgin HW (1998) Murine cytomegalovirus infection inhibits IFN gamma-induced MHC class II expression on macrophages: the role of type I interferon. Virology 241: 331–344

18. Hengel H, Flohr T, Hammerling GJ, Koszinowski UH, Momburg F (1996) Human cytomegalovirus inhibits peptide translocation into the endoplasmic reticulum for MHC class I assembly. J Gen Virol 77: 2287–2296

19. Hengel H, Koopmann JO, Flohr T, Muranyi W, Goulmy E, Hammerling GJ, Koszinowski UH, Momburg F (1997) A viral ER-resident glycoprotein inactivates the MHC-encoded peptide transporter. Immunity 6: 623–632

20. Hengel H, Koszinowski U (1997) Interference with antigen processing by viruses. Curr Opin Immunol 9: 470–476

21. Hill A, Jugovic P, York I, Russ G, Bennink J, Yewdell J, Ploegh H, Johnson D (1995) Herpes simplex virus turns off the TAP to evade host immunity. Nature 374: 411–415

22. Huang Z, Vafai A, Lee J, Mahalingam R, Hayward AR (1992) Specific lysis of targets expressing varicella-zoster virus gpI or gpIV by CD4$^+$ human T-cell clones. J Virol 66: 2664–2669

23. Jones T, Hanson L, Sun L, Slater J, Stenberg R, Campbell A (1995) Multiple independent loci within the human cytomegalovirus unique short region down-regulate expression of major histocompatibility complex class I heavy chains. J Virol 69: 4830–4841

24. Jones T, Sun L (1997) Human cytomegalovirus US2 destabilizes major histocompatibility complex class I heavy chains. J Virol 71: 2970–2979

25. Jones TR, Wiertz EJ, Sun L, Fish KN, Nelson JA, Ploegh HL (1996) Human cytomegalovirus US3 impairs transport and maturation of major histocompatibility complex class I heavy chains. proc Natl Acad Sci USA 93: 11327–11333

26. Kleijnen M, Huppa J, Lucin P, Mukherjee S, Farrell H, Campbell A, Koszinowski U (1997) A mouse cytomegalovirus glycoprotein, gp34, forms a complex with folded class I MHC molecules in the ER which is not retained but is transported to the cell surface. EMBO J 16: 685–694

27. Leonard GT, GC Sen (1997) Restoration of interferon responses of adenovirus E1A-expressing HT1080 cell lines by overexpression of p48 protein. J Virol 71: 5095–5101

28. Leopardi R, Ilonen J, Mattila L, Salmi AA (1993) Effect of measles virus infection on MHC class II expression and antigen presentation in human monocytes. Cell Immunol 147: 388–396

29. Miller DM, Rahill BM, Boss JM, Lairmore MD, Durbin JE, Waldman JW, Sedmak DD (1998) Human cytomegalovirus inhibits major histocompatibility complex class II expression by disruption of the Jak/Stat pathway. J Exp Med 187: 675–683

30. Machold R, Wiertz E, Jones T, Ploegh H (1997) The HCMV gene products US11 and US2 differ in their ability to attack allelic forms of murine major histocompatibility complex (MHC) class I heavy chains. J Exp Med 185: 363–366

31. Moffat JF, Stein MD, Kaneshima H, Arvin AM (1995) Tropism of varicella-zoster virus for human CD4$^+$ and CD8$^+$ T lymphocytes and epidermal cells in SCID-hu mice. J Virol 69: 5236–5242

32. Neefjes JJ, Momburg F (1993) Cell biology of antigen presentation. Curr Opin Immunol 5: 27–34

33. Petit AJ, Terpstra FG, Miedema F (1987) Human immunodeficiency virus infection down-regulates HLA class II expression and expression and induces differentiation in promonocytic U937 cells. J Clin Invest 79: 1883–1889

34. Ploegh HL (1998) Viral strategies of immune evasion. Science 280: 248–253

35. Reusch U, Muranyi W, Lucin P, Burgert H, Hengel H, Koszinowski U (1999) A cytomegalovirus glycoprotein re-routes MHC class I complexes to lysosomes for degradation. EMBO 18: 1081–1091

36. Scholz M, Hamann A, Blaheta RA, Auth MK, Encke A, Markus BH (1992) Cytomegalovirus- and interferon-related effects on human endothelial cells. Cytomegalovirus infection reduces upregulation of HLA class II antigen expression after treatment with interferon-γ. Hum Immunol 35: 230–238

37. Sedmak DD, Guglielmo AM, Knight DA, Birmingham DJ, Huang EH, Waldman WJ (1994) Cytomegalovirus inhibits major histocompatibility class II expression on infected endothelial cells [see comments]. Am J Pathol 144: 683–692

38. Sharp M, Terada K, Wilson A, Nader S, Kinchington PE, Ruyechan WT, Hay J, Arvin AM (1992) Kinetics and viral protein specificity of the cytotoxic T lymphocyte response in healthy adults immunized with live attenuated varicella vaccine. J Infect Dis 165: 852–858

39. Tomazin R, Hill AB, Jugovic P, York I, van Endert P, Ploegh HL, Andrews DW, Johnson DC (1996) Stable binding of the herpes simplex virus ICP47 protein to the peptide binding site of TAP. EMBO J 15: 3256–3266

40. Wiertz EJ, Jones TR, Sun L, Bogyo M, Geuze HJ, Ploegh HL (1996) The human cytomegalovirus US11 gene product dislocates MHC class I heavy chains from the endoplasmic reticulum to the cytosol. Cell 84: 769–779

41. York IA, Roop C, Andrews DW, Riddell SR, Graham FL, Johnson DC (1994) A cytosolic herpes simplex virus protein inhibits antigen presentation to CD8[+] T lymphocytes. Cell 77: 525–535

42. Zhang Y, Cosyns M, Levin MJ, Hayward AR (1994) Cytokine production in varicella zoster virus-stimulated limiting dilution lymphocyte cultures. Clin Exp Immunol 98: 128–133

43. Ziegler H, Thale R, Lucin P, Muranyi W, Flohr T, Hengel H, Farrell H, Rawlinson W, Koszinowski UH (1997) A mouse cytomegalovirus glycoprotein retains MHC class I complexes in the ERGIC/cis-Golgi compartments. Immunity 6: 57–66

Authors' address: Dr. A. M. Arvin, G-311, Stanford University School of Medicine, 300 Pasteur Drive, Stanford, CA 94305, U.S.A.

Pathway of viral spread in herpes zoster: detection of the protein encoded by open reading frame 63 of varicella-zoster virus in biopsy specimens

T. Iwasaki[1], R. Muraki[2], T. Kasahara[1], Y. Sato[1], T. Sata[1], and T. Kurata[1]

[1]Department of Pathology, National Institute of Infectious Diseases, Tokyo, Japan
[2]Department of Dermatology, Kasumigaura National Hospital, Tsuchiura, Japan

Summary. Reactivation of varicella-zoster virus (VZV) in the dorsal root or trigeminal ganglia causes herpes zoster. The pathway of viral spread from the ganglia to the skin and also within the skin is not yet completely understood. Histological studies have revealed that each skin lesion in herpes zoster progresses sequentially through the stages of erythema, vesicles, pustules and finally ulceration. An immunohistochemical study of the early skin lesions of herpes zoster demonstrated a high incidence of hair follicle involvement and the main localization of the virus at the isthmus. This evidence suggests that VZV initially spreads from the ganglia through myelinated nerves, which predominantly end around the isthmus of hair follicles. To further investigate the viral spread within the skin, we analyzed the sequential appearance of the immediate early proteins encoded by ORF 63 of VZV (IE63), using an anti-IE63 antibody raised by immunization of rabbits with a recombinant protein. This antibody could detect IE63 in a western blot analysis of infected cells and also in immunohistochemical analysis of the skin lesions of herpes zoster. IE63 initially appeared in the nuclei of the follicular epithelial cells and basal or parabasal epidermal cells. Later, the nuclei and cytoplasm of cells in the epidermis and hair follicles became positive. IE63 remained in the virus-infected cells even during their degeneration. When we examined the hair follicles in the early erythematous lesions, cells positive for IE63 were predominantly distributed around the isthmus. In addition, some lymphocytes around the blood vessels were also positive for IE63, but these cells were seldom positive for the structural antigen. Thus, these observations suggest that VZV arriving through myelinated nerves infects not only permissive cells, but also non-permissive cells in the involved skin of herpes zoster.

Introduction

Herpes zoster (shingles) is an acute disorder characterized by unilateral radicular pain and vesicular eruptions involving 1–3 dermatomes and is caused by reactivated varicella-zoster virus (VZV) persisting in the dorsal root or cranial sensory ganglia after primary VZV infection, chicken pox [1, 2]. The mechanisms of neural spread of reactivated VZV to the skin are poorly understood. Characteristic cells having eosinophilic intranuclear inclusions in the corresponding ganglia are observed with inflammatory changes and hemorrhagic necrosis [3–5]. Replication of VZV in the ganglia during herpes zoster was also confirmed by electron microscopy, indirect immunofluorescence [6–8], and isolation of the infectious viruses [9].

In this report, we review our findings in skin specimens obtained from patients with herpes zoster for investigation of intradermal and neural spread of VZV in this disease. In addition, we report on our study of the in vivo expression of the immediate early protein encoded by ORF 63 of VZV and its histological localization, in order to identify the VZV-infected cells in the herpes zoster.

Sequential changing of skin eruptions in herpes zoster

Macroscopically, the skin eruptions of herpes zoster vary in appearance and show sequential changes (Fig. 1). Each skin lesion starts as an erythematous lesion, goes through the stages of micro- and macroscopic vesicles and pustules, and finally becomes ulcerated and covered with a crust. Some erythematous and microvesicular lesions heal without undergoing further changes.

Microscopic observations have supported this sequence of changes of herpes zoster lesions [10]. Erythematous lesions are not associated with any marked epidermal changes, except for the formation of a perivascular halo and slightly enlarged nuclei of cells. Capillaries and venules in the dermis become dilated with perivascular cuffing by lymphocytes (Fig. 2). Thereafter, microvesicular changes are observed in the epidermis. Some epidermal cells show ballooning with acantholysis and occasionally contain intranuclear inclusions of the full or Cowdry A type. Occasionally, multinucleated giant cells having intranuclear

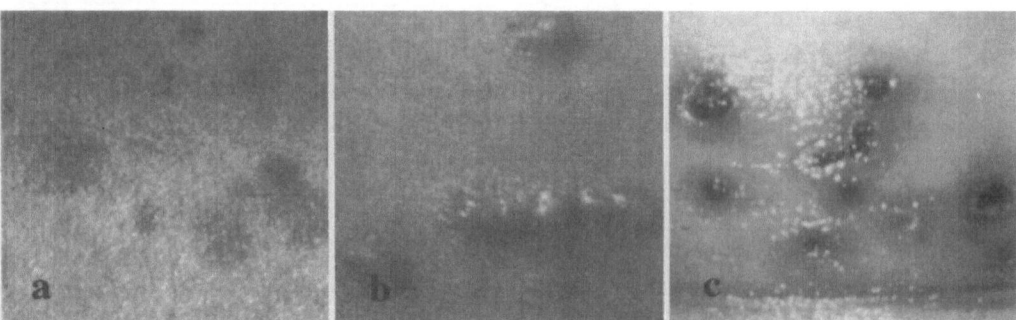

Fig. 1. Macroscopic appearances of the skin lesions of herpes zoster. **a** Erythematous macules. **b** Macroscopic vesicles. **c** Ulceration with crusts

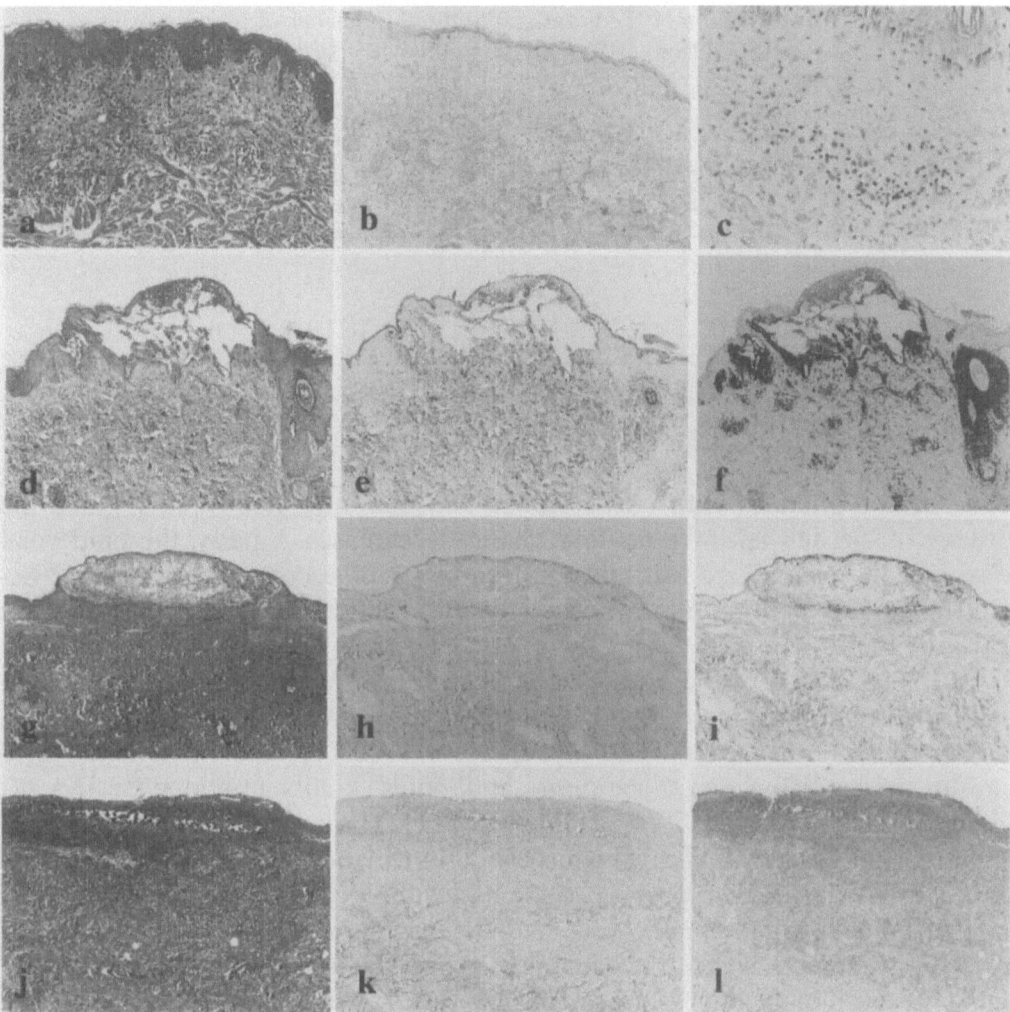

Fig. 2. Sequential changes of skin lesions of herpes zoster in histology. Skin lesions are histologically divided into four: The erythematous (**a**, **b**, **c**), vesicular (**d**, **e**, **f**), pustular (**g**, **h**, **i**) and ulcerating (**j**, **k**, **l**) stages. A skin biopsy specimen obtained from the erythematous stage shows perivascular cuffing of mononuclear cells (**a**, hematoxylin-eosin). No marked epidermal changes are observed and the structural antigens of VZV are not detectable (**h**, immunohistochemistry for NP/GP). But IE63 is detected in the nuclei of infiltrating lympho-cytes (**c**). A vesicular skin lesion (**d**, hematoxylin-eosin) shows massive antigen distribution of VZV both in the epidermis and dermis (**e**, NP/GP and f, IE63). In the pustular stage (**g**, hematoxylin-eosin) the antigen-positive cells are decreased in number especially in der-mis (**h**, NP/GP and **i**, IE63). Finally, an ulcerating skin lesion is covered by necrotic tissue (**j**, hematoxylin-eosin) and few antigen-positive cells are detected (**k**, NP/GP and **l**, IE63)

inclusions are detected among these lesions. Later, the intraepidermal vesicles become enlarged (Fig. 2). Moderate perivascular infiltration, predominantly by neutrophils, is observed in the upper dermis. In late vesicular lesions, leukoclastic vasculitis is observed. Vesicles covered with necrotic or degenerated epidermal

Table 1. Localization of VZV antigen in the skin lesions of herpes zoster
(modified from [10], with the publisher's permission)

	Stage			
	erythematous $n = 7$	vesicular $n = 14$	pustular $n = 9$	ulcerative $n = 7$
Epidermis	1	14	9	1[a]
Hair follicle	1	6	5	1[a]
Dermis	0	14	6	0

[a]Only in the superficial layers

cells become pustules with intravesicular infiltration of neutrophils. Then the severity of the dermal inflammatory changes decreases. Finally, the epidermal cells over the lesions are lost and the lesions become covered by necrotic tissue. Lymphocytes, plasma cells and histiocytes infiltrate the dermis beneath the necrotic tissue.

The localization of the structural proteins of VZV was immunohistochemically investigated in these biopsy specimens [10]. We could detect a 32 kDa capsid protein (NP) and a 64 kDa glycoprotein 3 (GP) of VZV in formalin-fixed paraffin tissues using two monoclonal antibodies, kindly provided by Drs. K. Yamanishi and T. Okuno, Osaka University [11]. These structural proteins were detected in the epidermis and follicular epithelium in the late erythematous lesions (Table 1). These antigen-positive cells had intranuclear inclusion bodies and showed ballooning and acantholysis, with the formation of microvesicles (Fig. 2). After the formation of the macroscopic vesicles, the number of antigen-positive cells, increased in the dermis. These positive cells were fibroblast, macrophages, and vascular endothelium. In the pustular stage, the number of antigen-positive cells in the dermis decreased again and these cells disappeared completely in the ulcerative stage [10]. In the ulcerative stage, only a few degenerating or necrotic cells of epithelial origin were positive for VZV.

Hair follicle infection in herpes zoster

To clarify the pathway of viral spread from the ganglia to the skin, we analyzed the early skin lesions by histological and immunohistochemical methods [12]. VZV-infected cells were more frequently detected in the hair follicles than in the epidermis (Table 2). Within the involved hair follicles, the infected cells were localized within the isthmus (12/12), less frequently in the stem (8/10) and infundibulum (6/10), and were not observed at all in the bulb (Table 3).

The high frequency of isthmus involvement in herpes zoster suggests that VZV spreads first to the area of the skin innervated by myelinated nerves, which predominantly end around the isthmus of hair follicles.

Viral spread in herpes zoster 113

Table 2. Comparison of viral involvement in the hair follicles and epidermis in the erythematous lesions of herpes zoster (from [12] with the publisher's permission)

Viral involvement	No of specimens ($n = 16$)
None	4
Only in hair follicle	10
Only in epidermis	2
Both in hair follicle and epidermis	0

Viral involvement was determined by histological or immunohistochemical detection of the virus-infected cells and/or the GP/NP antigens of VZV, respectively

Table 3. Localization of VZV infection within the hair follicle in the erythematous lesions of herpes zoster (from [12] with the publisher's permission)

Subdivision	Observed number	VZV involvement
Infundibulum	10	6 (60%)
Isthmus	12	12 (100%)
Stem	10	8 (80%)
Hair bulb	9	0 (0%)

Immediate early protein encoded by ORF 63 of VZV

Among the 71 open reading frames (ORFs) of VZV, ORFs 4, 61, 62 and 63 have been shown to encode immediate early (IE) proteins, which are homologues to ICPs 27, 0, 4 and 22, respectively, of herpes simplex virus-1 [13]. Among these IE proteins, the ORF 63 product of VZV (IE63) has been demonstrated to be abundantly expressed during the latent and early infective period of VZV in vivo [14]. To identify the early-infected cells in herpes zoster, we investigated the sequential appearance of IE63 in the skin lesions by immunohistochemical analysis.

Preparation of IE63 and rabbit antiserum

We produced a recombinant IE63 protein as a glutathione S-transferase fusion protein in *E. coli*. Briefly, the sequence of VZV ORF63 was amplified using the oligonucleotide primer pair of 5'-GGG ATC CAG GAC ATG TTT TGC ACC T as the forward primer (nt.110572–110591 of GenBank X04370) and 5'-AGA ATT CTA TTT ATT TAT AAA GAC T as the reverse primer (nt.111416–111440). This selected sequence for protein expression is similar to that for the recombinant GST fusion protein prepared by Debrus et al. [14]. Restriction enzyme recognition sites (*Bam*HI or *Eco*RI) were inserted into both primers. We used the S*st*I B fragment of VZV, kindly provided by Dr. Y. Hayakawa as the template for polymerase chain reaction (PCR). After amplification of ORF63 sequence of VZV by

114 T. Iwasaki et al.

PCR, the amplicon, 869 bp in length, was digested with *Bam*HI and *Eco*RI and inserted into the *Bam*HI and *Eco*RI sites of pGEX-3X (Amersham Pharmacia, Buckinghamshire, UK). The recombinant plasmids were confirmed by sequence analysis. After transfection, the GST-IE63 fusion protein was induced in DH5 of *E. coli* by addition of isopropyl 1-thio-β-**D**-galactoside and purified by glutathione sepharose beads. Then, we immunized two New Zealand rabbits with 100 μg of the purified recombinant protein five times, monthly, by subcutaneous injection with complete (at first immunization) or with incomplete Freund's adjuvants (at booster immunization).

IE63 recognized by rabbit antiserum

Rabbit anti-IE63 antiserum was characterized by immunoblotting of VZV-infected and uninfected human embryonic lung fibroblast (HEL) cells. HEL cells with or without VZV infection were lyzed in 2 × SDS sample buffer with 2-mercapto-ethanol. The samples were electrophoresed in 10% SDS polyacrylamide gel and transferred to a PVDF membrane. The membrane was allowed to react with this antiserum. After washing in PBS, alkaline phosphatase-conjugated anti-rabbit IgG (ALI0405, Biosource International, CA) was added. The alkaline phosphatase activity was developed in a mixture solution of NBT and BCIP.

Although the amino acid sequence of this ORF predicts a 30.8 kDa acidic protein, three bands, 32 kDa and 62/64 kDa in size, were recognized in the samples of HEL cells infected by VZV, but no bands were recognized in those of uninfected HEL cells (Fig. 3). The 32 kDa band is compatible with a previous report [15], but 62/64 kDa doublets have not been reported so far. The presence of a 47 or 45 kDa band was reported in VZV-infected human melanoma cells or Vero cells, by Kinchington et al. [15] and Debrus et al. [14], respectively. IE63 has not been well characterized biochemically except for its phosphorylation [16]. Further biochemical analyses will be required to reveal the nature of these doublets of IE63.

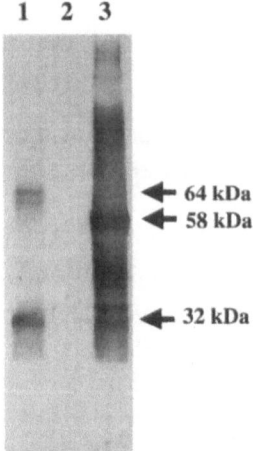

Fig. 3. Western blot analysis of VZV-infected cells by anti-IE63 rabbit serum. A major 32 kDa and two minor 62 and 64 kDa doublet bands are observed in Vero cells infected by VZV (*1*). No bands are detected in Vero cells without VZV infection (*2*). *3* is a sample of expressed IE63 as a GST-fusion proteins in *E. coli* (DH5)

Histological localization of IE63 in biopsy specimens obtained from cases of herpes zoster

The distribution of IE63 in the skin lesions was analyzed on serial paraffin sections of formalin-fixed biopsy specimens by the method used for staining the nucleocapsid/glycoprotein of VZV [10, 12]. In brief, after deparaffinization, the sections were treated with 0.25% trypsin (Difco, Detroit, MI) in PBS (v/v) containing 0.02% calcium chloride (w/v) at 37 °C for 30 min and immersed in 0.3% hydrogen peroxide prepared in methanol (v/v) for 30 min. Then, the sections were incubated initially with normal goat serum for 30 min and with rabbit anti-IE63 overnight at 4 °C. Biotinylated anti-rabbit IgG (E0432, DAKO, CA) was applied for 45 min at 37 °C, followed by the application of peroxidase conjugated streptavidin (P0397, DAKO) for 45 min at 37 °C. The peroxidase reaction was developed in 0.05M Tris buffer (pH 7.6) with 0.02% diaminobenzidine (DAKO) and 0.015% hydrogen peroxide. Nuclei were counterstained with 2% methyl green or hematoxylin.

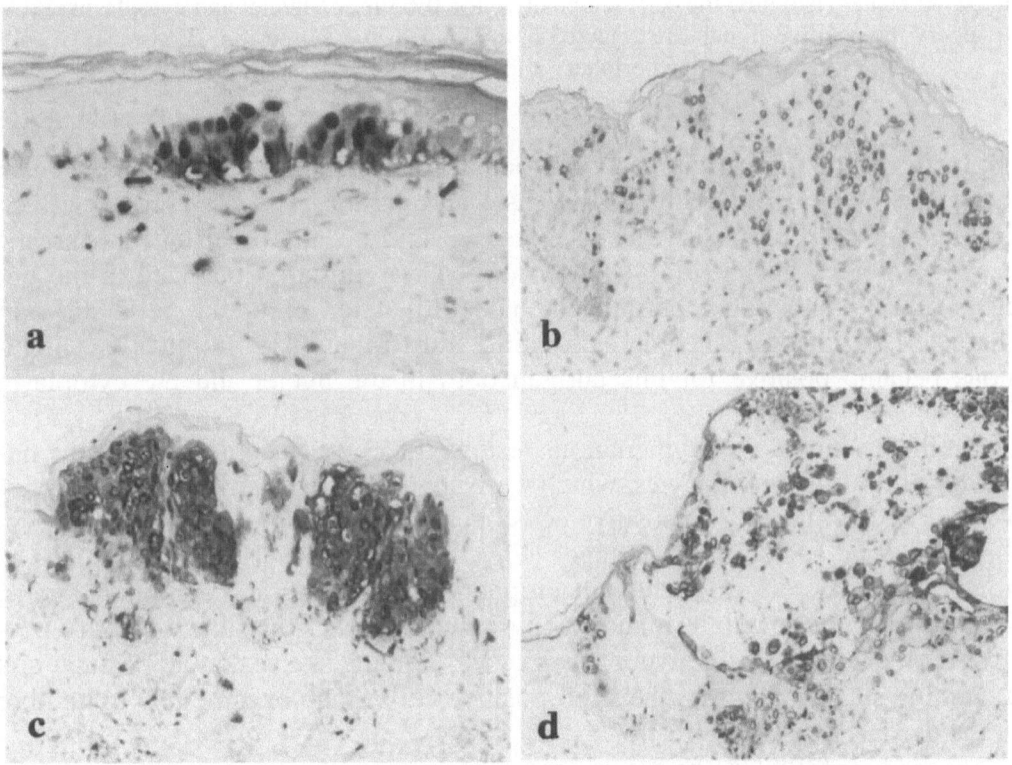

Fig. 4. Immunohistochemistry of sequential expression of IE63 in the epidermis. At the early erythematous stage, IE63 appears initially in the nuclei of epidermal cells in basal and parabasal layers (**a**). Some epidermal cells are also positive for IE63 in nuclei. At the late erythematous stage, the number of IE63-positive cells is increased (**b**). In the epidermis forming a microvesicle IE63 is localized in the nuclei and cytoplasm of epidermal cells (**c**). In the periphery, IE63 is detected only in the nuclei. In the pustular stage, degenerating VZV-infected cells in a pustule are positive of IE63 (**d**)

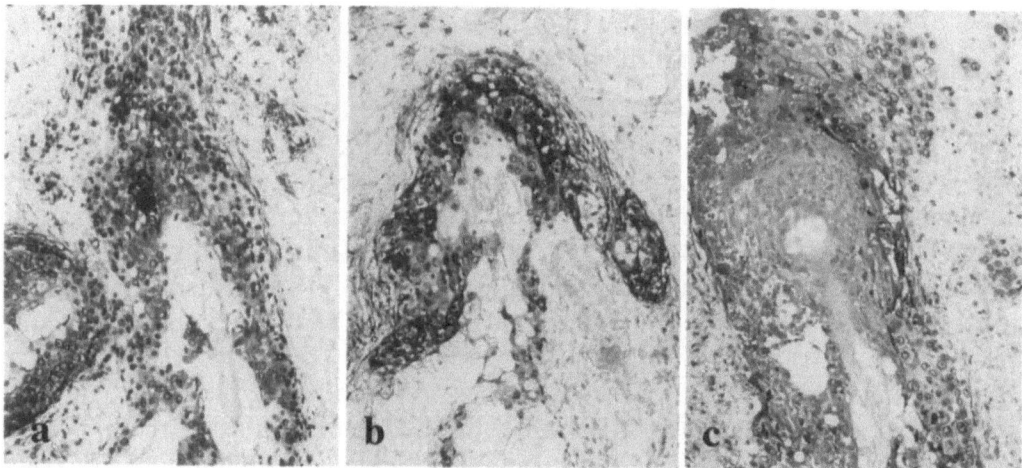

Fig. 5. Immunohistochemical detection of IE63 in the isthmus of hair follicles. In early lesions, IE63-positive cells are predominantly localized in the isthmus and sebaceous gland (**a, b**). At center (isthmus) the cells are positive for IE63 in the nuclei and cytoplasm and in periphery the positive signals are detected in the nuclei. In a late lesion, IE63-positive cells localized in the periphery and decreased in the isthmus (**c**)

In the epidermis in the early erythematous lesions, IE63 was observed in the nuclei of epidermal cells in the basal and parabasal layers (Fig. 4). Later, the number of IE63-positive cells increased and became distributed from the basal to spinous layers (Fig. 4). The antigen was initially localized in the nucleus of cells showing a perinuclear halo. Still later, epithelial cells showing characteristic changes such as ballooning and formation of intranuclear inclusions became positive for this antigen both in the nuclei and the cytoplasm (Fig. 4).

In the dermis of the erythematous lesions, IE63 was also detected in the nuclei of infiltrating lymphocytes, which were negative for the structural proteins of VZV (Fig. 2). These IE63-positive cells became more numerous in the late erythematous and early vesicular stages (Figs. 2 and 4). The IE63-positive cells of non-lymphocytic origin, such as macrophages, fibroblasts and endothelial cells, showed ballooning with formation of intranuclear inclusions, and were also positive for the nucleocapsid/glycoprotein of VZV. In the late vesicular, pustular and ulcerating stages, however, the IE63-positive cells again became few in number in the dermis.

In the hair follicles, the pattern of distribution of IE63 was similar to that in the epidermis (Fig. 5). However, it was difficult to find early follicular lesions in the biopsy specimens. Even in the erythematous stage, IE63-positive cells were rather numerous. The cells positive in the nucleus and cytoplasm for IE63 were recognized as being in the isthmus and stem of hair follicles in the examined biopsy specimens obtained from early skin lesions (Fig. 5). The cells only positive in the nucleus were recognized surrounding these fully developed infected cells.

Discussion

We have characterized the VZV-infected cells in the skin lesions of herpes zoster and clarified that VZV replication initially starts in the epithelial cells of the epidermis and hair follicles. Subsequently, VZV infected the dermal cells, but soon thereafter, during the late vesicular and pustular stages, the number of infected cells in the dermis again decreased in number. Epithelial cells infected with VZV disappeared in the pustular and ulcerating stages. The reason for the early disappearance of infected cells in the dermis compared with those in the epidermis has not been clarified. One possibility is that the host immune system attacks the infected cells in the dermis more effectively than it does those in the epidermis and follicular epithelium. The histological differences in the skin lesions between chicken pox (primary infection) and herpes zoster (reactivation) must be clarified to define the role of memory immune cells to attack and destroy VZV-infected.

The previous study on VZV-infected cells in the early skin lesions of herpes zoster suggested that the virus spreads from the dorsal root ganglia through the myelinated nerves, the endings of which surrounds the isthmus of hair follicles. The present immunohistochemical study of IE63 distribution in these lesions confirmed the early infection of the hair follicles.

This study first demonstrated the sequential changes of IE63 localization in infected cells during a proliferative VZV infection. It appears initially in the nucleus, followed by localization both in the nucleus and, abundantly, in the cytoplasm. Incorporation of IE63 into the viral particles, probably in the tegument [15], can be explained by the presence of IE63 in the cytoplasm. The function of IE63 other than as an immediate early protein in viral replication remains to be clarified.

Using an antibody that recognized the IE protein, we succeeded in detecting virus-infected cells in the early skin lesions, in which antibodies recognizing the structural protein of VZV failed to recognize VZV-infected cells. Surprisingly, the cells positive for IE63 were lymphocytes infiltrating around the vessels in the dermis. Although we could not find IE63-positive lymphocytes within the vascular lumen, the VZV genome has been shown to be present in the mononuclear cells of the peripheral blood (PBMC) obtained from patients with herpes zoster by DNA hybridization [17] and PCR analyses [18]. Mainka et al. reported that the transcription not only of IE63, but also of ORF68 encoding glycoprotein E, was detectable in these by isothermal-transcription-based nucleic acid amplification [18]. In our immunohistochemical study, these lymphocytes were seldom positive for antigens of the structural proteins of VZV. This suggests that lymphocytic infection of VZV is usually abortive or partially lytic; however, lymphocyte-mediated immune response through recognition of IE63 by T cells is important in herpes zoster [19].

Acknowledgements

We gratefully acknowledge Drs. Koichi Yamanishi and Toshiomi Okuno, Department of Virology, Osaka University School of Medicine by providing the monoclonal antibodies for

VZV and Dr. Yasuhiko Hayakawa, Sendai Municipal Institute of Public Health, Japan for supplying the cloned fragments of VZV DNA.

References

1. Hyman RW, Ecker JR, Tenser RB (1983) Varicella-zoster virus RNA in human trigeminal ganglia. Lancet II: 814–816
2. Gilden DH, Vafai A, Shatram Y, Becker Y, Devlin M, Wellish M (1983) Varicella-zoster virus DNA in human sensory ganglia. Nature 306: 478–480
3. Cheatham WJ (1953) The relation of heretofore unreported lesions to pathogenesis of herpes zoster. Am J Pathol 29: 401–411
4. Ghatak NR, Zimmermann HM (1973) Spinal ganglion in herpes zoster: a light and electron microscopic study. Arch Pathol 95: 411–415
5. Head H, Champbell AW (1900) The pathology of herpes zoster and its bearing on sensory localization. Brain 23: 353–523
6. Esiri MM, Tomlinson AH (1972) Herpes zoster: demonstration of virus in trigeminal nerve and ganglion by immunofluorescence and electron microscopy. J Neurol Sci 15: 35–48
7. Shibuta H, Ishikawa T, Aoyama Y, Kurata K, Matsumoto M (1974) Varicella virus isolation from spinal ganglion. Arch Ges Virusforsch 45: 382–385
8. Nagashima K, Nakazawa M, Endo H, Kurata T, Aoyama Y (1975) Pathology of the human spinal ganglia in varicella-zoster virus infection. Acta Neuropathol 33: 105–117
9. Bastian FO, Rabson AS, Yee CL, Tralka TS (1974) Herpesvirus varicellae: Isolated from human dorsal root ganglia. Arch Pathol 97: 331–333
10. Muraki R, Baba T, Iwasaki T, Sata T, Kurata T (1992) Immunohistochemical study of skin lesions in herpes zoster. Virchows Arch (A) 420: 71–76
11. Okuno T, Yamanishi K, Shiraki K, Takahashi M (1983) Synthesis and processing of glycoproteins of varicella-zoster virus (VZV) as studied with monoclonal antibodies to VZV antigens. Virology 129: 357–368
12. Muraki R, Iwasaki T, Sata T, Sato Y, Kurata T (1996) Hair follicle involvement in herpes zoster: pathway of viral spread from ganglia to skin. Virchows Arch 428: 275–280
13. Cohen JI, Straus SE (1996) Varicella-zoster virus and its replication. In: Fields BN, Knipe DM, Howley PM, Chanock RM, Melnick JL, Monath TP, Roizman B, Straus SE (eds) Fields Virology. Lippincott-Raven, Philadelphia, pp 2525–2545
14. Debrus S, Sadzot-Delvaux C, Nikkels AF, Piette J, Rentier B (1995) Varicella-zoster virus gene 63 encodes an immediate-early protein that is abundantly expressed during latency. J Virol 69: 3240–3245
15. Kinchington PR, Bookey D, Turse SE (1995) The transcriptional regulatory proteins encoded by varicella-zoster virus open reading frames (ORFs) 4 and 63, but not ORF61, are associated with purified virus particles. J Virol 69: 4274–4282
16. Stevenson D, Xue M, Hay J, Ruyechan WT (1996) Phosphorylation and nuclear localization of the varicella-zoster virus gene 63 protein. J Virol 70: 658–662
17. Vonsover A, Leventon-Kriss S, Langer A, Smetana Z, Zaizov R, Potaznick D, Cohen IJ, Gotlieb-Stematsky T (1987) Detection of varicella-zoster virus in lymphocytes by DNA hybridization. J Med Virol 21: 57–66
18. Mainka C, Fuß B, Geiger H, Höfelmayr, Wolff MH (1998) Characterization of viremia at different stages of varicella-zoster virus infection. J Med Virol 56: 91–98

19. Sadzot-Delvaux C, Kinchington PR, Debrus S, Rentier B, Arvin AM (1997) Recognition of the latency-associated immediate early protein IE63 of varicella-zoster virus by human memory T lymphocytes. J Immunol 159: 2802–2806

Authors' address: Dr. T. Iwasaki, Department of Pathology, National Institute of Infectious Diseases, Toyama 1-23-1, Shinjuku-ku, Tokyo 162-8640, Japan.

Vaccination against cytomegalovirus

S. A. Plotkin

Medical and Scientific Advisor, Pasteur Mérieux Connaught Emeritus Professor
of Pediatrics, University of Pennsylvania Emeritus Professor,
Wistar Institute, Doylestown, Pennsylvania, U.S.A.

Summary. Like varicella zoster virus (VZV), human cytomegalovirus (HCMV) causes disease after both primary and recurrent infections. The former is more serious, particularly in pregnant women, who may transmit the virus to their offspring, with a high risk of mental retardation and deafness. Various experimental vaccines are in development, ranging from live, attenuated HCMV, subunit envelope glycoprotein, poxvirus vectors with CMV genes inserted, and plasmid DANN.

Active immunization. Attenuated live vaccines

The first attempts to immunize against human cytomegalovirus (CMV) were pursued in the 1970s by two groups, one at St. George's Hospital in London led by Stephen Elek [15, 36], and the other my own group at Children's Hospital and the Wistar Institute of Philadelphia [45]. The British group started with the prototype laboratory-adapted AD-169 strain, whereas we started with a new isolate from a congenitally infected infant named Towne. Both vaccines were based on classic live-attenuated viruses which induced seroconversion in vaccinees by subcutaneous injection, with slight and tolerable local and systemic symptoms. The development of the Towne strain as a vaccine started in the mid-1970s, based on the simple idea that passage of the virus in cell cultures would eventually succeed in attenuating it for humans. The only practical cell substrate was human fibroblast cell strains, but in fact this was also the logical substrate, in that our laboratory had used it to grow attenuated rubella, attenuated polio, and inactivated rabies viruses [43]. Differences in in vitro markers between high-passage and low-passage Towne virus were identified [70], but because human CMV will not infect other species, precluding the use of any heterologous animal models, only clinical trials could tell us whether the virus had been attenuated.

Accordingly, the first clinical trial of the Towne virus (125th passage in MRC-5 cells) was performed in healthy male young adult volunteers [44]. The

Table 1. Properties of Towne strain live attenuated cytomegalovirus vaccine

- ■ Safety
  - No systemic symptoms or clinical laboratory abnormalities
  - Not isolated from those recently vaccinated
  - No evidence for latency after immunosuppression
  - Injection site reaction common without systemic side effects
  - No depression of cell-mediated immunity or alteration of CD4/CD8 ratio

- ■ Immunogenicity
  - Regularly elicits humoral response, including neutralizing antibody
  - Induces lasting lymphocyte proliferative responses
  - Induces HLA-restricted cytotoxicity

- ■ Efficacy
  - In seronegative renal allograft recipients, does not prevent infection but does decrease frequency and mitigate severity of CMV-induced disease
  - In normal volunteers, protects against low-dose wild-type challenge
  - In mothers of children excreting CMV, did not prevent infection

observations, later confirmed in many volunteers of both sexes [32], were that the subcutaneous injection was well tolerated, but that local reactions consisting of erythema and swelling occurred at about a week post inoculation. Subsequent studies suggested that the reaction was the equivalent of a delayed-type hypersensitivity reaction at the site of CMV antigen deposition, related to the presence of inactivated antigen inoculated with the live virus [23].

With regard to immunogenicity, antibody responses evolved as they do after natural infection, although to lower titers. In particular, neutralizing antibodies regularly appeared by 4 weeks post-vaccination. Lymphocyte proliferation to CMV antigen, surrogate for CD4+ cell activation, became evident 1 to 2 weeks post-vaccination. In addition, CD8+ Cytotoxic T cells (CTL) were detected when tested, although, unlike after natural infection, it tended to fade by 1 year post-vaccination [3, 54].

Additional findings with the Towne strain are summarized in Table 1. Attempts to recover virus from the blood, saliva and urine of vaccinees were uniformly negative. Biopsies done at the site of injection 1–2 days later occasionally recovered CMV, although it was not clear whether this was inoculated virus or new virus.

Because CMV disease is a major problem in renal transplant patients, randomized, placebo-controlled prospective studies were organized to test the ability of the Towne strain vaccine to prevent disease, and also to test its safety in immunocompromised persons [48, 49]. In these trials, prospective renal transplant recipients received either Towne vaccine or placebo, and then were followed post-transplant for infection and disease. Severity of disease was evaluated blindly according to a scoring system. As had been previously determined [58, 62], the high-risk group for disease was comprised of recipients seronegative at study

Table 2. Comparative results of 3 blinded trials of towne vaccine in seronegative renal transplant patients who received kidneys from seropositive donors

Trial	n	Rate of all CMV disease[a]		Rate of severe CMV disease[a]		Reduction of severe disease in vaccinated compared with placebo
		V	P	V	P	
Pennsylvania	67	39%	55%	6%	35%	84%
Minnesota	35	33%	43%	5% (10%)[b]	36%	87%
Multicentric	61	38%	59%	0%	17%	100%
All	163	37%	54%	3%	29%	89%

[a] *P* Patients given placebo; *V* patients given vaccine

[b] A 10% rate was reported in the original publication, but this includes 1 case that occurred subsequent to pancreatic transplant after a renal transplant free of CMV disease. Without that case, the incidence was 5%

entry who received kidneys from seropositive donors. Superinfection with a donor CMV strain is known to occur in seropositive recipients but the result is usually less serious [13, 62].

Three randomized, controlled, double-blind studies, one of which was multicentric, were performed in renal transplant patients to determine the protection afforded by Towne vaccine, all giving similar results (Table 2) [7, 46, 49]. Vaccination with Towne did not prevent patients from being infected with CMV, but considerably modified the severity of their disease. In fact, a reduction of about 85% in severe disease was achieved, similar to the protection provided by prior natural infection.

Moreover, the large DNA genome of CMV permitted the unequivocal identification of excreted virus strains by restriction-endonuclease migration patterns, or RFLPs, as they are known today [47]. The excreted viruses were compared with the vaccine virus, and none was identical. In addition, when Towne vaccine was given to seronegative recipients of kidneys from seronegative donors, they did not excrete any CMV [48]. Thus, in these cases, the vaccine virus failed to reactivate in the immunosuppressed host, suggesting that it produced limited replication and had not become latent.

To test the efficacy of vaccine in a group who would act as surrogates for pregnant women, two trials were done in healthy volunteers. One study used a challenge with a low tissue-culture passage CMV, called Toledo [50]. At the fifth cell culture passage, subcutaneous injection of Toledo induced symptoms of heterophile-negative infectious mononucleosis, with fever, increases in hepatic enzymes and decreases in platelet counts becoming evident 4 to 5 weeks post-injection, and resolving spontaneously. The volunteers for this study were Catholic priests in a seminary, to whom I owe a debt of gratitude. Each was tested to identify those who were naturally seropositive or seronegative. Some of the latter received Towne vaccine about ~1 year before challenge, whereas others

Table 3. Challenge dose of subcutaneously administered Toledo strain (low passage) CMV required to infect or cause disease in 50% of subjects in different groups

Group	No. of pfu producing indicated outcome in 50%		
	Infection	Laboratory abnormalities	Disease
Seronegative	10	≤ 10	≤ 10
Naturally seropositive	1000	~ 500	1,000
Vaccinated	100	100	> 100

were retained as controls. The results are summarized in Table 3. Unvaccinated seronegative volunteers became ill after injection of 10 or 100 plaque-forming units (PFU) of challenge virus. Naturally seropositive volunteers resisted 10 or 100 PFU, but about one-half became ill after challenge with 1000 PFU of Toledo. Vaccinated volunteers also resisted low doses, but were somewhat more susceptible to infection than the naturally immune group.

The second challenge was more naturalistic: adult women with young children attending day-care centers were given either Towne vaccine or placebo and then followed for virological and serological evidence of CMV infection [5]. It was previously known that in day-care centers children become infected from one another, and that transmission to their mothers occurs in about ~40% of cases [39]. The results of this study were disappointing in that the infection rate did not differ between vaccinated and control mothers, although the resistance of naturally infected women was striking.

Thus, it appeared that although the Towne strain vaccine does give an immunogenic, mild infection, the immunity induced is not as substantial as that after natural infection. Wang et al. [67] hypothesized that the difference lay in the titer of neutralizing antibodies, which was lower after vaccination than after natural infections and they are studying higher dosage regimens. In any case, the results showed that the immune response to Towne, either humoral or cellular, would have to be improved.

Improved versions of Towne

The results with Towne had been mainly obtained with pools of virus manufactured by SmithKline Beecham Biologicals (Rixensart, Belgium) or Merck Research Laboratories (West Point, PA). However, a third pool had been prepared by a contract with Microbiological Associates (McLean, VA), and Adler et al. [6] found that the latter pool was more immunogenic with respect to antibody production. Consequently, a new efficacy trial has been launched in mothers of children in day-care centers to see if the higher titers of antibodies elicited by this new strain translates to greater efficacy.

An interesting, genetic approach has been taken by workers at Aviron (Mountainview, CA), a biotech company specializing in live vaccine development. They noticed that in the Towne vaccine there is a virus variant with a deletion of 13 kb of DNA when compared to the Toledo wild virus [12]. Sequencing of the deletion, which is in the unique long b' region of the genome, revealed 19 open reading frames presumably coding for proteins. They hypothesized that the genes missing from Towne vaccine contribute to the virulence and immunogenicity of CMV and that restoration of some or all of those genes might result in a more immunogenic vaccine. Accordingly, the Aviron group constructed hybrids between Towne and Toledo, to restore the deletion, and also to incorporate other defined parts of the Toledo genome as replacements for Towne sequences [35]. These hybrids will be tested in humans to see whether induction of immunity has been enhanced. If successful, this genetic approach will yield a mutant strain that might duplicate natural infection. Clinical trials will determine whether the right balance of immunogenicity and safety can be achieved.

Subunit glycoprotein B

As described above, the epitopes that induce neutralizing antibodies are present on the envelope glycoproteins of CMV, of which there are three major groups. The best studied is the gB protein, which is responsible for at least half of neutralizing antibodies in the serum of naturally infected individuals [9, 53]. There are a number of reasons for thinking that gB-elicited responses might provide immunity: these are summarized in Table 4.

Early on, our laboratory demonstrated that the envelope from CMV virions is immunogenic in humans and that affinity-purified gB could induce neutralizing antibodies [21]. However, in view of the poor growth of CMV in cell culture, it was impractical to consider virus-derived gB as the source of a vaccine. Scientists at the Chiron Corp. (Emeryville, CA) took up the challenge of developing a practical source of gB using genetic engineering techniques. They inserted the gene for gB into Chinese Hamster ovary cells, obtaining a stably transfected cell line [59]. Furthermore, they truncated the 3' end of the gene, in order to obtain better secretion of gB.

Table 4. Data supporting the utility of Glycoprotein gB in protection against CMV

- Sera from naturally infected individuals readily precipitate purified gB
- Seronegative volunteers immunized with Towne strain, or experimentally infected with a low-passive strain (Toledo) all developed antibody to gB
- Adsorption of sera with gB removes most neutralizing antibodies
- Monoclonal antibodies that recognize gB are able to neutralize laboratory CMV strains as well as clinical isolates
- Mice and guinea pig inoculated with vaccinia recombinant expressing gB develop neutralizing antibody and are protected
- Seronegative and seropositive volunteers were immunized intramuscularly with immunoaffinity purified gB; the naturally immune individuals developed booster responses in neutralizing antibody and lymphocyte proliferation, and the seronegative volunteers developed neutralizing antibodies and cellular responses, although several injections were necessary

Table 5. Antibody response to three or four doses of subunit CMV glycoprotein B with MF59 adjuvant (Chiron) administered to adults on a schedule of 0, 1, 6 and 18 months

Bleeding schedule (months)	Time	Neutralizing titers (GMT)
7	Post Dose 3	100
18	Pre Dose 4	30
19	Post Dose 4	110
24	Post Dose 4	65
30	Post Dose 4	60

Having once obtained sufficient quantities of purified gB, the Chiron workers then had to deal with the relatively poor immunogenicity of viral glycoproteins. To overcome that problem, they applied a new proprietary oil-in-water adjuvant called MF-59, which had already been shown to significantly enhance the immunogenicity of herpes simplex glycoproteins and HIV envelope glycoproteins [11]. The results [1, 29, 34, 40] (Frey et al., pers. comm., Pass et al., pers. comm.) with CMV gB, summarized in Table 5, showed a definite neutralizing antibody response after three doses of antigen with adjuvant [34]. The optimal dose of gB was determined to be 30 mcg, and the optimal schedule 0, 30 and 180 days. The antibodies elicited by gB neutralized all primary CMV isolates tested [29].

Clinical studies were performed in healthy adults [40] and in toddlers [34], confirming good immunogenicity of the adjuvanted gB vaccine in both age groups. However, it was noted that the levels of neutralizing antibodies fell rapidly during the six months after the third dose. Therefore, a fourth dose was administered, resulting in an anamnestic response that seemed to have greater persistence, although extended data are not yet available [40].

The induction of high titers of neutralizing antibodies by the subunit gB prompted the organization of an efficacy study at the University of Alabama, under the direction of Robert Pass (pers. comm.). The population in which the vaccine is being tested consists of mothers from a low socioeconomic background who are known to have a high rate of intercurrent CMV infection. These women will be given either gB vaccine or placebo after their first pregnancy, and then followed until they become pregnant again, when their second child will be tested at birth for evidence of intrauterine CMV infection.

Glycoprotein H

Glycoprotein H (gH) is another candidate for a subunit vaccine, as it carries neutralizing epitopes [64]. Moreover, antibodies to gH may reduce the chance of contracting retinitis [56], perhaps because they act on viral release and thus affect cell-to-cell transfer of CMV [26]. Although at least one vaccine company has expressed interest in the use of gH, there are no published data as yet.

Canarypox recombinants

The poxviruses, of which vaccinia is also the prototype, are known for their ability to induce cellular immune responses [8]. However, vaccinia virus can replicate unchecked in immunocompromised hosts, and is associated with other serious adverse reactions [25]. In 1982, Paoletti [38], Moss and their coworkers [31] published data showing that foreign sequences of DNA could be inserted into the very large genomes of poxviruses, and that the proteins coded for by the sequences would be expressed. The poxviruses could thus serve as vectors to express antigens of interest to vaccinators.

Paoletti then went on to explore the use of avian poxviruses as safe alternatives to vaccinia, whereas Moss [61] concentrated on an attenuated mutant vaccinia. Paoletti [63] identified the canarypox virus as a useful vector, because it can be grown in avian cells but does not replicate in mammalian cells, providing a large safety factor for humans. Nevertheless, despite its inability to produce infectious virions in mammals, canarypox does transcribe and translate information from "early" genes (those expressed before DNA synthesis), and if DNA sequences encoding antigenic proteins of vaccine interest are inserted downstream of early promotors, their genetic information will also be expressed.

The earliest demonstration of the possible utility of canarypox for human vaccination was with a recombinant carrying the glycoprotein gene of rabies virus [19, 42]. The canarypox-rabies virus induced neutralizing antibodies sufficient to protect animals against challenge and in humans also produced titers thought to be protective. Subsequent studies with other canarypox recombinants coding for less immunogenic proteins showed that antibody induction required multiple doses, but that cellular responses were excellent after two to three doses. Moreover, two doses powerfully primed the immune system for B-cell responses to subsequently administered protein vaccines. This "prime-boost" approach is currently in Phase 2 tests of candidate HIV vaccines, and is considered to be promising [17].

After animal tests had shown good immunogenicity, a canarypox-gB recombinant was constructed and tested in humans [20]. Unfortunately, the volunteer studies showed only weak antibody responses after two to three doses of the recombinant [4]. To simulate a protein boost, a study was then done in which volunteers were primed either with canarypox-gB or canarypox-rabies, followed by an injection of the Towne attenuated virus which produces gB as part of its replication [4]. The results (Table 6) showed a definite priming of the antibody response to gB and of the neutralizing antibody response. Thus, a basis was established for testing canarypox priming and subunit gB boost. Under the aegis of National Institute for Allergy and Infectious Diseases (NIAID), Pasteur Mérieux Connaught and Chiron are collaborating to test the prime-boost protocol in adults at the Cincinnati Children's Hospital, under the direction of David Bernstein. This study has begun, and final results are expected in mid-1999.

Meanwhile, another aspect of canarypox recombinants is being explored: their ability to induce cellular immunity, particularly CTL. Work by Stanley Riddell and colleagues in Seattle [68], McLaughlin-Taylor et al. in Duarte, CA [33],

Table 6. CMV antibody responses one month after two doses of ALVAC-CMV (gB) or ALVAC-Rabies followed by a booster subunit gB with chiron

Priming	GMT ELISA[a]	GMT Neut[b]
ALVAC-gB	38,802 (100-409,600)	186 (25-645)
ALVAC-RG	1,345 (100-64,000)	27 (15-116)

[a] Against gB antigen
[b] Reciprocal of plaque neutralization titers

Table 7. Positive CMV-CTL tests (>10% specific lysis) after canarypox-pp65 vaccination at months 0, 1, 3 and 6

Group	Month of sampling				
	0	3 (Post 2 doses)	4 (Post 3 doses)	5	7 (Post 4 doses)
Seronegative vaccinees	0/6	6/6	6/6	6/6	6/6
Seronegative placebo	0/3	N.D.	0/3	N.D.	0/3
Natural seropositives	3/3	1/1	3/3	N.D.	2/2

N.D. Not done

and by my colleagues Eva Gonczol and Klara Berencsi at the Wistar Institute in Philadelphia [10] has confirmed that the principal targets of CMV-specific CTL are the pp65 matrix antigen and the IE1 (Immediate-Early 1) nonstructural antigen. A canarypox-pp65 recombinant was constructed and tested in clinical trial (Table 7). Vaccinees uniformly developed antibody responses to pp65, but more importantly, they also showed strong CTL responses to cells displaying pp65 [24]. These responses should provide protection against CMV disease, as suggested by the results of Riddell's studies of exogenously administered T lymphocytes sensitized primarily to pp65 [28, 66]. Now the task is to combine in one vaccine regimen all those antigens that might provide protection.

DNA plasmids

This approach, described in the literature using the term "naked DNA", stems from the serendipitous discovery that injection of bacterial plasmids containing inserted genes can elicit immune responses to the proteins corresponding to the genes [65, 69]. Although there are several major safety issues yet to be resolved, DNA plasmids are attractive immunogens, and are particularly adept at inducing cellular immunity. Therefore, the first studies of plasmids containing CMV genes have been undertaken with pp65 as the encoded antigen. My colleagues [16] and workers at the City of Hope Hospital (Duarte, CA) [37] each succeeded in demonstrating excellent responses in mice to pp65 plasmid DNA, indicating that this strategy may have a place in CMV vaccine development.

Vaccination based on peptide induction of CMV-specific CTL is also under investigation [14].

How can efficacy be tested?

The issue of efficacy testing is not a simple one for investigators or for vaccine manufacturers. In general, the Phase 3 efficacy evaluation is the most expensive part of vaccine development, and if one arrives at that stage without prior evidence for efficacy, it becomes a Las Vegas all or nothing gamble. In the case of CMV the question is whether one has to show prevention of congenital disease, or whether there is a surrogate indicator for that endpoint. Logically it can be argued that prevention of infection of normal seronegative women by contact with CMV-infected children should protect their fetuses from congenital infection during pregnancy. However, pregnancy is immunosuppressive, and the counter-argument could be made that it is important to show protection of fetuses from infection, when vaccinated mothers are exposed to CMV. This may be all the more true when one takes into consideration that 1–2% of seropositive mothers will give birth to CMV-excreting infants, even if those infants seem to be protected against disease [60]. Moreover, serially infected siblings may be born to an immune mother, although not with the same prognosis [27].

A trial to show prevention of congenital infection in a situation where the prevalence of that infection is 1% is feasible, and could be done if required. However, because of the late appearance of clinical signs of CMV infection, to show prevention of congenital CMV disease would require a much larger trial involving tens of thousands of women, lasting long enough for infants to grow to school age, and the systematic testing of those infants for intelligence and for hearing ability, a task likely to daunt any vaccine manufacturer.

Table 8. Trials to demonstrate efficacy of a CMV vaccine

Vaccinated population	Endpoint	Incidence in placebo group	Number subjects[a]	
			(1 year follow-up)	(2 years follow-up)
Mother of children in day care	Infection	25%[b]	368	184
Pre-pregnant women	Fetal infection	1%	9190	4595
Pre-pregnant women	Fetal infection	5%[c]	1838	919
Pre-pregnant women	Fetal disease	0.1%[b]	91900	45950

[a]Assuming vaccine efficacy is 80%, confidence limit about 50%, $\alpha = 0.05$, $\alpha = 0.8$. Number is total of vaccine and placebo groups

[b]Conservative estimate

[c]High risk adolescents in lower socio-economic group

As Table 8 shows, the efficacy of a CMV vaccine could be tested at several levels, and preliminary evidence of efficacy could be obtained in relatively small Phase 2 trials, to avoid conducting very expensive Phase 3 trials on ineffective vaccines. A first level of evidence of efficacy would be the prevention of infection in mothers whose children acquire CMV in day care. If those studies showed the vaccine to be protective, the next step would be to vaccinate a large number of women and to follow them during subsequent pregnancies, in parallel with a suitable control group of women. Their newborn infants would be tested for infection by virus isolation or PCR [18, 30], from urine and saliva. Protection against fetal infection should suffice to prove efficacy against fetal disease. Although it is conceivable that a vaccine might prevent disease but not fetal infection, that hypothesis is probably too difficult to test.

When a CMV vaccine is licensed, how would we use it?

From a public health viewpoint, the primary objective of a CMV vaccine is to prevent infection during pregnancy. This would imply the vaccination of all adolescent girls, or older women who intend to become pregnant. An adolescent vaccination date has been discussed at the ACIP, and for girls a routine vaccination visit between 11 and 13 years would be ideal. An economic evaluation of vaccination with the live, attenuated vaccine concluded that its use would be cost-effective [52]. Of course, the duration of immunity induced by a licensed vaccine will be critical. If immunity is short-lived, vaccination might be advised just before pregnancy is contemplated; if immunity is of medium duration (5 years), regular boosters could be given throughout the child-bearing years; and if immunity is long-lived, both adolescent and pediatric vaccination could be considered. If immunity after vaccination is as long-lasting as after natural infection in infancy, the thrust of development would change towards incorporation of CMV antigens into combination vaccines given during the first year of life.

In view of declining seropositivity in young women living in industrialized countries, now about 50% [22, 55], serologic testing before vaccination would not seem to be indicated, and even less so for adolescent girls.

No doubt a licensed CMV vaccine could be given to solid organ (kidney, liver, heart, lung) transplant recipients before surgery, and also might be used to vaccinate bone marrow donors so that they could serve as sources of transfused T cells active against CMV.

Acknowledgements

This paper is an abbreviated version of an article: Plotkin, S. A. (1999) Vaccination against cytomegalovirus, the changeling demon. Pediatr Infect Dis J 18: 313–326.

References

1. Adler SP (1996) Current prospects for immunization against cytomegaloviral disease. Infect Agents Dis 5: 29–35

2. Adler SP (1986) Molecular epidemiology of cytomegalovirus: evidence for viral transmission to parents from children infected at a day care center. Pediatr Infect Dis 5: 315–318

3. Adler SP, Hempfling SH, Starr SE, Plotkin SA, Riddell S (1998) Safety and immunogenicity of the Towne strain cytomegalovirus vaccine. Pediatr Infect Dis J 17: 200–206

4. Adler SP, Plotkin SA, Gonczol E, Cadoz M, Meric C, Wang JB, Dellamonica P, Best AM, Zahradnik J, Pincus S, Berencsi K, Cox WI, Gyulai Z (1999) A canarypox vector expressing cytomegalovirus (CMV) glycoprotein B primes for antibody responses to a live attenuated CMV vaccine (Towne). J Infect Dis 180: 843–846

5. Adler SP, Starr SE, Plotkin SA, Hempfling SH, Buis J, Manning ML, Best AM (1995) Immunity induced by primary human cytomegalovirus infection protects against secondary infection among women of childbearing age. J Infect Dis 171: 26–32

6. Adler SP, Hempfling SH, Starr SE, Plotkin SA, Riddell S (1998) Safety and immunogenicity of the Towne strain cytomegalovirus vaccine. Pediatr Infect Dis 17: 200–206

7. Balfour HH (1991) Prevention of cytomegalovirus disease in renal allograft recipients. Scand J Infect Dis 78: 88–93

8. Bennink JR, Yewdell JW, Smith GL, Moller C, Moss B (1984) Recombinant vaccinia virus primes and stimulates influenza haemagglutinin-specific cytotoxic T cells. Nature 311: 578–579

9. Britt WJ, Mach M (1996) Human cytomegalovirus glycoproteins. Intervirology 39: 401–412

10. Burian K, Endresz V, Gyulai Z (1999) Prevalence of the IE1-exon4-specific CTL in naturally seropositive healthy individuals. [Abstract] 7th International Cytomegalovirus Workshop, Brighton, UK, March 7–9

11. Burke RL (1991) Development of a Herpes simplex virus subunit glycoprotein vaccine for prophylactic and therapeutic use. J Infect Dis 13: S906–S911

12. Cha TA, Tom E, Kemble GW (1996) Human cytomegalovirus clinical isolates carry at least 19 genes not found in laboratory strains. J Virol 70: 78–83

13. Chou SW (1986) Acquisition of donor strains of cytomegalovirus by renal-transplant recipients. N Engl J Med 314: 1418–1423

14. Diamond DJ, York J, Sun JY, Wright CL, Forman SJ (1997) Development of a candidate HLA A*0201 restricted peptide-based vaccine against human cytomegalovirus infection. Blood 90: 1751–1767

15. Elek SD, Stern H (1974) Development of a vaccine against mental retardation caused by cytomegalovirus infection in utero. Lancet 1: 1–5

16. Endresz V, Kari L, Berencsi K, Kari C, Gyulai Z, Jeney C, Pincus S, Rodeck U, Meric C, Plotkin SA, Gonczol E (1999) Induction of human cytomegalovirus (HCMV)-glycoprotein B (gB)-specific neutralizing antibody and phosphorprotein 65 (pp65)-specific cytotoxic T lymphocyte responses by naked DNA immunization. Vaccine 17: 50–58

17. Excler JL, Plotkin SA (1997) The prime-boost concept applied to HIV preventive vaccines. AIDS 11: S127–S137

18. Fox JC, Kidd IM, Griffiths PD, Sweny P, Emery VC (1995) Longitudinal analysis of cytomegalovirus load in renal transplant recipients using a quantitative polymerase chain reaction: correlation with disease. J Gen Virol 1995, 76: 309–319

19. Fries LF, Tartaglia J, Taylor J, Kauffman EK, Meignier B, Paoletti E, Plotkin S (1996) Human safety and immunogenicity of a canarypox-rabies glycoprotein recombinant vaccine: an alternative poxvirus vector system. Vaccine 14: 428–434

20. Gonczol E, Berencsi K, Pincus S, Endresz V, Meric C, Paoletti E, Plotkin SA (1995) Preclinical evaluation of an ALVAC (canarypox)-human cytomegalovirus glycoprotein B vaccine candidate. Vaccine 13: 1080–1085

21. Gonczol E, Ianocone J, Ho W, Starr S, Meignier B, Plotkin S (1990) Isolated gA/gB glycoprotein complex of human cytomegalovirus envelope induces humoral and cellular immune-responses in human volunteers. Vaccine 8: 130–136

22. Gratacap-Cavallier B, Bosson JL, Morand P, Dutertre N, Chanzy B, Jouk PS, Vandekerckhove C, Cart-Lamy P, Seigneurin JM (1998) Cytomegalovirus seroprevalence in French pregnant women. Eur J Epidemiol 14: 147–152

23. Gupta R, Gonczol E, Manning ML, Starr S, Johnson B, Murphy GF, Plotkin SA (1993) Delayed type hypersensitivity to human cytomegalovirus. J Med Virol 39: 109–17

24. Gyulai Z, Pincus S, Cox B (1999) Canarypox-CMV-pp65 recombinant immunization of seronegative subjects elicits pp65 specific CTL precursors with a frequency comparable to pp65 specific frequency of naturally seropositive individuals. [Abstract] 7th International Cytomegalovirus Workshop, Brighton, UK, March 7–9

25. Henderson DA, Moss B (1999) Smallpox. In: Plotkin SA, Orenstein WA (eds) Vaccines. W. B. Saunders, Philadelphia, pp 74–97

26. Keay S, Baldwin B (1991) Anti-idiotype antibodies that mimic gp86 of human cytomegalovirus inhibit viral fusion but not attachment. J Virol 65: 5124–5128

27. Krech U, Konjajev Z, Jung M (1971) Congenital cytomegalovirus infection in siblings from consecutive pregnancies. Helv Paediatr Acta 26: 355–362

28. Li CR, Greenberg PD, Gilbert MJ, Goodrich JM, Riddell SR (1994) Recovery of HLA-restricted cytomegalovirus (CMV)-specific T-cell responses after allogeneic bone marrow transplant: correlation with CMV disease and effect of ganciclovir prophylaxis. Blood 83: 1971–1979

29. Liu H, Chou S, Sekulovich R, Duliege A-M, Burke RL (1997) A CMV glycoprotein gB subunit vaccine elicits cross neutralizing antibodies that cross neutralize clinical isolates. Abstract 43. Sixth International Cytomegalovirus Workshop. March 7–9, Perdido Beach, AL

30. Lucht E, Brytting M, Bjerregaard L (1998) Shedding of cytomegalovirus and herpesviruses 6, 7, and 8 in saliva of human immunodeficiency virus type 1-infected patients and healthy controls. Clin Infect Dis 27: 137–141

31. Magno A, Smith GL, Moss B (1982) Vaccinia virus. A selectable eukaryotic cloning and expression vector. Proc Natl Acad Sci USA 79: 7415–7419

32. Marshall GS, Plotkin SA (1994) Progress toward developing a cytomegalovirus vaccine. Infect Dis Clinics N Am 4: 283–298

33. McLaughlin-Taylor E, Pande H, Forman SJ, Tanamachi B, Li CR, Zaia JA, Greenberg PD, Riddell SR (1994) Identification of the major late human cytomegalovirus matrix protein pp65 as a target antigen for CD8+ virus-specific cytotoxic T lymphocytes. J Med Virol 43: 103–110

34. Mitchell D, Holmes SJ, Burke RL, Sekulovich R, Tripathi M, Doyle M, Duliege AM (1997) Immunogenicity of a recombinant human cytomegalovirus (CMV) gB vaccine in toddlers. Sixth International Cytomegalovirus Workshop. March 5–9 (Abstract no. 50)

35. Mocarski ES Jr, Kemble GW (1996) Recombinant cytomegaloviruses for study of replication and pathogenesis. Intervirology 39: 320–330

36. Neff BJ, Weibel RE, Buynak EB, McAllen AA, Hillman MR (1979) Clinical and laboratory studies of live c cytomegalovirus vaccine Ad-169. Proc Soc Exp Biol Med 160: 32–37

37. Pande H, Campo K, Tanamachi B, Forman SJ, Zaia JA (1998) Direct DNA immunization of mice with plasmid DNA encoding the tegument protein pp65 (ppUL83) of human

cytomegalovirus induces high levels of circulating antibody to the encoded protein. Scand J Infect Dis [Suppl] 99: 117–120

38. Panicali D, Paoletti E (1982) Construction of poxviruses as cloning vectores: Insertion of the thymidine kinase gene from herpes simplex virus into the DNA of infectious vaccinia virus. Proc Natl Acad Sci USA 79: 4927–4931

39. Pass RF, Hutto C, Ricks R, Cloud GA (1986) Increased rate of cytomegalovirus infection among parents of children attending day-care centers. N Engl J Med 314: 1414–1418

40. Pass R, Duliège AM, Sekulovich R, Boppana S, Hirabayashi S, Britt WJ, Burke RL (1997) Antibody response to a fourth dose of CMV gB vaccine in healthy adults. Abstract 157. Sixth International Cytomegalovirus Workshop. March 7–9: Perdido Beach, AL

41. Plotkin SA (1999) Cytomegalovirus. In: Plotkin SA, Orenstein WA (eds) Vaccines. Saunders, Philadelphia, pp 903–908

42. Plotkin SA, Cadoz M, Meignier B, Meric C, Leroy O, Excler JL, Tartaglia J, Paoletti E, Gonczol E, Chappuis G (1995) The safety and use of canarypox vectored vaccines. Dev Biol Stand 84: 165–170

43. Plotkin SA, Eagle H, Hayflick L, Ikic D, Koprowski H, Perkins F (1969) Serially cultured animal cells for the preparation of viral vaccines. Science 165: 1282

44. Plotkin SA, Farquhar J, Horberger E (1976) Clinical trials of immunization with the Towne 125 strain of human cytomegalovirus. J Infect Dis 134: 470–475

45. Plotkin SA, Furukawa T, Zygraich N, Huygelen C (1975) Candidate cytomegalovirus strain for human vaccination. Infect Immun 12: 521–527

46. Plotkin SA, Higgins R, Kurtz JB (1994) Multicenter trial of Towne strain attenuated virus vaccine in seronegative renal transplant recipients. Transplantation 58: 1176–1178

47. Plotkin SA, Huang ES (1985) Cytomegalovirus vaccine virus (Towne strain) does not induce latency. J Infect Dis 152: 395–397

48. Plotkin SA, Smiley ML, Friedman HM, Starr SE, Fleisher GR, Wlodaver C, Dafoe DC, Friedman AD, Grossman RA, Barker CF (1984) Towne-vaccine-induced prevention of cytomegalovirus disease after renal transplants. Lancet 1: 528–530

49. Plotkin SA, Starr SE, Friedman HM, Brayman K, Harris S, Jackson S, Tustin NB, Grossman R, Dafoe D, Barker C (1991) Effect of Towne live virus vaccine on cytomegalovirus disease after renal transplant. A controlled trial. Ann Intern Med 114: 525–531

50. Plotkin SA, Starr SE, Friedman HM, Gonczol E, Weibel RE (1989) Protective effects of Towne cytomegalovirus vaccine against low-passage cytomegalovirus administered as a challenge. J Infect Dis 159: 860–865

51. Plotkin SA, Starr SE, Friedman HM, Gonczol E, Brayman K (1990) Vaccines for the prevention of human cytomegalovirus infection. Rev Infect Dis [Suppl] 7: S827–S838

52. Porath A, McNutt RA, Smiley LM, Weigle KA (1990) Effectiveness and cost benefit of a proposed live cytomegalovirus vaccine in the prevention of congenital disease. Rev Infect Dis 12: 31–40

53. Qadri I, Navarro D, Paz P, Pereira L (1992) Assembly of conformation-dependent neutralizing domains on glycoprotein B of human cytomegalovirus. J Gen Virol 73: 2913–2921

54. Quinnan GV Jr, Delery MRA, Frederick WR, Epstein JS, Manischewitz JF, Jackson L, Ramsey KM, Mittal K, Plotkin SA (1984) Comparative virulence and immunogenicity of the Towne strain and a nonattenuated strain of cytomegalovirus. Ann Intern Med 101: 478–483

55. Rosenthal S, Stanberry LR, Biro FM, Slaoui M, Francotte M, Koutsoukos M, Hayes M, Bernstein DI (1997) Seroprevalence of Herpes simplex virus types 1 and 2 and cytomegalovirus in adolescents. Clin Infect Dis 24: 135–139

56. Rasmussen L, Morris S, Wolitz R, Dowling A, Fessell J, Holodniy M, Merigan TC (1994) Deficiency in antibody response to human cytomegalovirus glycoprotein gH in human immunodeficiency virus-infected patients at risk for cytomegalovirus retinitis. J Infect Dis 170: 673–677

57. Sinzger G, Jahn G (1996) Human cytomegalovirus cell tropism and pathogenesis. Intervirology 39: 302–319

58. Smiley ML, Wlodaver CG, Grossman RA, Barker CF, Perloff LJ, Tustin NB, Starr SE, Plotkin SA, Friedman HM (1985) The role of pretransplant immunity in protection. Transplantation 40: 157–161

59. Spaete RR (1991) A recombinant subunit vaccine approach to HCMV vaccine development. Transplant Proc 23: 90–96

60. Stagno S, Reynolds DW, Huang ES (1977) Congenital cytomegalovirus infection. N Engl J Med 296: 1254–1258

61. Sutter G, Wyatt LS, Foley P, Bennink JR, Moss B (1994) A recombinant vector derived from the host range-restricted and highly attenuated MVA strain of vaccinia virus stimulates protective immunity in mice to influenza virus. Vaccine 12: 1032–1040

62. Suwansirikul S, Rao N, Dowling JN, Ho M (1977) Primary and secondary cytomegalovirus infection. Arch Intern Med 137: 1036–1029

63. Taylor J, Meignier B, Tartaglia J, Languet B, VanderHoeven J, Franchini G, Trimarchi C, Paoletti E (1995) Biological and immunogenic properties of a canarypox-rabies recombinant, ALVAC-RG (vCP65) in non-avian species. Vaccine 13: 539–549

64. Urban M, Klein M, Britt WJ, Hassfurther E, Mach M (1996) Glycoprotein H of human cytomegalovirus is a major antigen for the neutralizing humoral immune response. J Gen Virol 77: 1537–1547

65. Ulmer JB, Donnelly JJ, Parker SE, Rhodes GH, Felgner PL, Dwarki VJ, Gromkowski SH, Deck RR, DeWitt CM, Friedman A (1993) Heterologous protection against influenza by injection of DNA encoding a viral protein. Science 259: 1745–1749

66. Walter EA, Greenberg PD, Gilbert MJ, Finch RJ, Watanabe KS, Thomas ED, Riddell SR (1982) Reconstruction of cellular immunity against cytomegalovirus in recipients of allogeneic bone marrow by transfer of T-cell clones from the donor. N Engl J Med 307: 7–13

67. Wang JB, Adler SP, Hempfling S, Burke RL, Duliege AM, Starr SE, Plotkin SA (1996) Mucosal antibodies to human cytomegalovirus glycoprotein B occur following both natural infection and immunization with human cytomegalovirus vaccines. J Infect Dis 174: 387–392

68. Wills MR, Carmichael AJ, Mynard K, Jin X, Weekes MP, Plachter B, Sissons JG (1996) The human cytotoxic T-lymphocyte (CTL) response to cytomegalovirus is dominated by structural protein pp65: frequency, specificity, and T-cell receptor usage of pp65-specific CTL. J Virol 70: 7569–7579

69. Wolff JA, Malone RW, Williams P, Chong W, Acsadi G, Jani A, Felgner PL (1990) Direct gene transfer into mouse muscle in vivo. Science 247: 1465–1468

70. Yamane Y, Furukawa T, Plotkin SA (1983) Supernatant virus release as a differentiating marker between low passage and vaccine strains of human cytomegalovirus. Vaccine 1: 25

Author's address: Dr. S. A. Plotkin, 4650 Wismer Road, Doylestown, PA 18901, U.S.A.

Varicella zoster virus in human and rat tissue specimens

P. W. Annunziato[1], O. Lungu[2], and C. Panagiotidis[3]

[1]Departments of Pediatrics and [2]Microbiology, College of Physicians and Surgeons,
Columbia University, New York, New York, U.S.A.
[3]Department of Pharmaceutical Sciences, Aristotle University, Thessaloniki, Greece

Summary. The limited supple of appropriate tissues for study has been an impediment to investigations of varicella zoster virus (VZV) latency. Human dorsal root ganglia (DRG) harboring latent virus are not plentiful and are not amenable to manipulation for studying the events surrounding the establishment, maintenance, and cessation of latency. An alternative to studies in human DRG is the rat model of latency, which appears to provide a reliable method of investigating VZV latency. Other alternatives include studies in other human tissues involved in VZV pathogenesis. In order to improve our understanding of the establishment and cessation of latency, we performed comparative immunohistochemical analysis of chickenpox and zoster skin lesions. This analysis revealed that during primary infection and reactivation productive VZV infection occurs in a variety of cell types and that the major VZV DNA binding protein, ORF29p, is present in peripheral axons early during the course of chickenpox. VZV latency was studied in the rat model by in situ hybridization and compared with similar studies performed in human DRG containing latent virus, confirming that VZV DNA persists in the same sites in DRG of the two species.

Introduction

Primary varicella zoster virus (VZV) infection causes chickenpox and latent infection in host dorsal root ganglia (DRG), placing the host at risk for virus reactivation, which may be clinically recognized as zoster. The events that facilitate the persistence of VZV in the neurons and satellite cells of the DRG remain poorly understood. Although it appears that the entire virus genome is present during latency [9, 14], only a portion of the genome is expressed at this stage [4–7, 15]. Proteins encoded by open reading frames (ORFs) expressed during latency accumulate in neurons but localize to atypical cellular compartments [13]. When VZV reactivates in DRG, the full lytic cycle cascade resumes and the proteins are found in locations that are typical of productive infection [13]. The factors that influence these events have not yet been fully elucidated.

Hindering further investigations of these phenomena is the lack of appropriate tissue that supports latent VZV infection. VZV is highly species-specific and naturally infects only humans and, rarely, some non-human primates [3]. The most reliable small animal model for the study of VZV latency is the rat model [16]. VZV appears to establish a latent infection in DRG of rats following subcutaneous inoculation and virus DNA persists for months at this site [1, 16]. Whether persistence of VZV in rat DRG results from the same events that allow persistence of VZV in human DRG is still to be determined. In the meantime, studies of the establishment, maintenance, and cessation of VZV latency are largely performed in specimens from the natural host.

Here we describe investigations aimed at elucidating the events involved in VZV latency and reactivation. In order to gain better insight into the establishment and cessation of latency, human skin biopsy specimens from patients with chickenpox and zoster were evaluated by immunohistochemistry [2]. VZV latency was studied in the rat model by in situ hybridization [1] and the findings were compared with similar studies performed in human DRG [12].

VZV infection in human skin biopsies

Comparative immunohistochemistry of skin biopsies from patients with chickenpox and zoster were performed using polyclonal antibodies generated against the protein products of immediate early (IE)63, ORF29, and ORF14 [2]. These proteins were chosen for this analysis to include a representative of the three kinetic classes of VZV genes. By analogy to the herpes simplex virus homologue, ICP8, ORF29 is an early gene that encodes a DNA binding protein [3]. ORF 14, a late gene, encodes the glycoprotein, gC [3]. IE63p and gC are virion components [11]. ORF29p is not a component of the virion [10] and its presence in a cell nucleus implies virus replication at that site .

Six cases of chickenpox and 8 cases of zoster were selected for immunohistochemical analyses (Table 1, Fig. 1) [1]. Five cases of Grover's disease, a

Table 1. Detection of VZV proteins in skin biopsies

		IE 63p			ORF 29p				gC		
Dx	+/N[a]	EP	EN	NE	EP	EN	NE	WBC[b]	EP	EN	NE
CP	6/6	4	3	0	6	4	2	4	4	5	3
Z	8/8	5	1	0	8	4	0	5	5	5	4

Tissues from patients with clinical and histopathological diagnoses (DX) of chickenpox (CP) or zoster (Z) were analyzed by immunohistochemistry for immediate early (IE 63p), early (ORF 29p) and late (gC) virus proteins in epithelial (EP), endothelial (EN), inflammatory cells expressing CD43 (WBC), or in dermal nerves (NE). The results are expressed as the absolute number of biopsies with detectable protein. Zero indicates absence of detectable protein

[a]The number of cases positive for any VZV protein/the number of cases examined

[b]2 chickenpox cases and 2 zoster cases with VZV-infected WBC were found to express CD68 and not CD3 or CD20. Reproduced with permission [2]

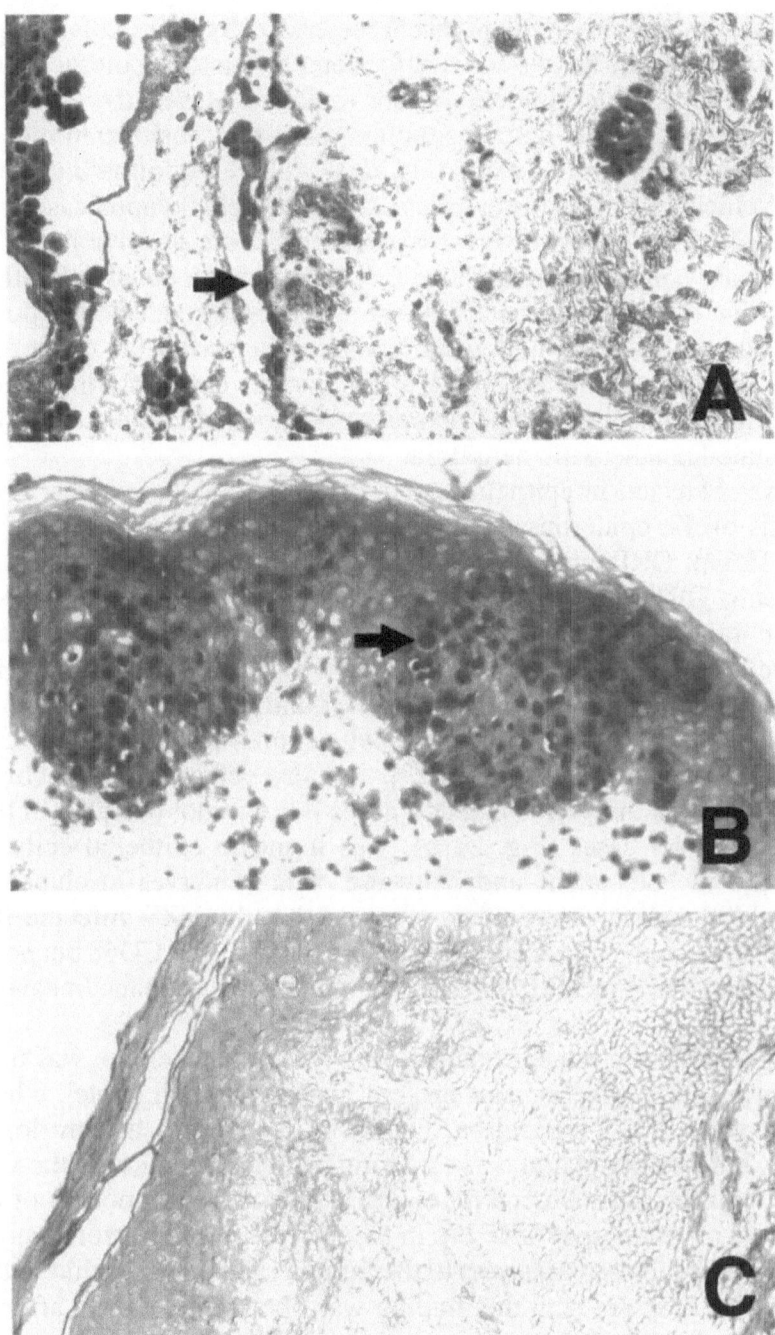

Fig. 1. Immunohistochemical detection of ORF29p in skin biopsies. Chickenpox (**A**), zoster (**B**) and Grover's disease (**C**) skin lesions were analyzed for ORF29p. The arrows indicate positive epithelial cells. The sections are shown at 100× magnification. Reproduced with permission [2]

P. W. Annunziato et al.

non-infectious dermatologic disorder characterized by a vesicular eruption that is distinguishable from herpes virus infections by histopathologic criteria, and 3 cases of herpes simplex infection were included as negative controls. None of the Grover's disease or herpes simplex cases had positive immunostaining (Fig. 1) [2]. Biopsies were obtained as early as 2 days following the onset of rash in both chickenpox and zoster cases. All of the 6 chickenpox cases were positive for ORF29p, 4 were positive for IE63p and 5 were positive for gC. Among the 8 zoster cases, all were positive for ORF29p, 5 were positive for IE63p, and 5 were positive for gC. One zoster patient subsequently developed postherpetic neuralgia. The biopsy specimen was obtained 3 days after the onset of the rash in this case. When compared with the other 7 zoster cases, no differences could be detected in the histopathologic or immunohistochemical features of the case that precipitated postherpetic neuralgia.

VZV was detected in epithelial cells, endothelial cells, nerves, and inflammatory cells of the epidermis and dermis in both chickenpox and zoster cases (Table 1). IE63p, ORF29p, and gC were detected in the expected intracellular compartments. ORF29p was found in cell nuclei (Figs. 1 and 2). gC was found in cell membranes and the cytoplasm (data not shown). IE63p was found in both the cell nucleus and cytoplasm (data not shown). IE63p was detected in epithelial cells and endothelial cells. ORF29p was found in epithelial cells, endothelial cells (Figs. 1 and 2), and inflammatory cells expressing the cell differentiation marker, CD43 (data not shown). ORF29p was also detected in the Schwann cells and axons of nerves in two chickenpox cases but was not detected in these cells in any of the zoster cases (Fig. 2). gC was found in epithelial cells, endothelial cells, and in both axons and Schwann cells of nerves in chickenpox and zoster. Double labeling experiments revealed that the VZV infected inflammatory cells in chickenpox and zoster cases express CD43 and CD68 but not CD20 or CD3, indicating that they are cells of the monocyte macrophage lineage (data not shown).

Finding ORF29p in the peripheral nerve during chickenpox was unexpected because during lytic infection this protein localizes to cell nuclei, where it participates in virus DNA replication. Unlike gC, which as an envelope protein presumably entered peripheral nerves during primary infection as the virus envelope fused with the axonal membrane, ORF29p is not a component of the virion [10]. De novo production of ORF29p in the nucleus of the neuron residing in the DRG, followed by axonal transport to the peripheral axon is not likely because at least one of the biopsies with this finding was obtained two days after the onset of the rash. It is widely thought that the skin is infected by hematogenous spread of VZV during chickenpox and that virus from the skin enters peripheral nerves to infect the neuron. A substantial time period would be necessary for the virus to travel from the skin to the DRG by retrograde axonal transport, a slow process of 200–400 mm/day [17]. Additional hours would lapse before early virus proteins such as ORF29p would be produced in the neuron cell body. A newly synthesized protein could then be detected in peripheral axons only after traveling the considerable distance back to the dermis by anterograde axonal transport

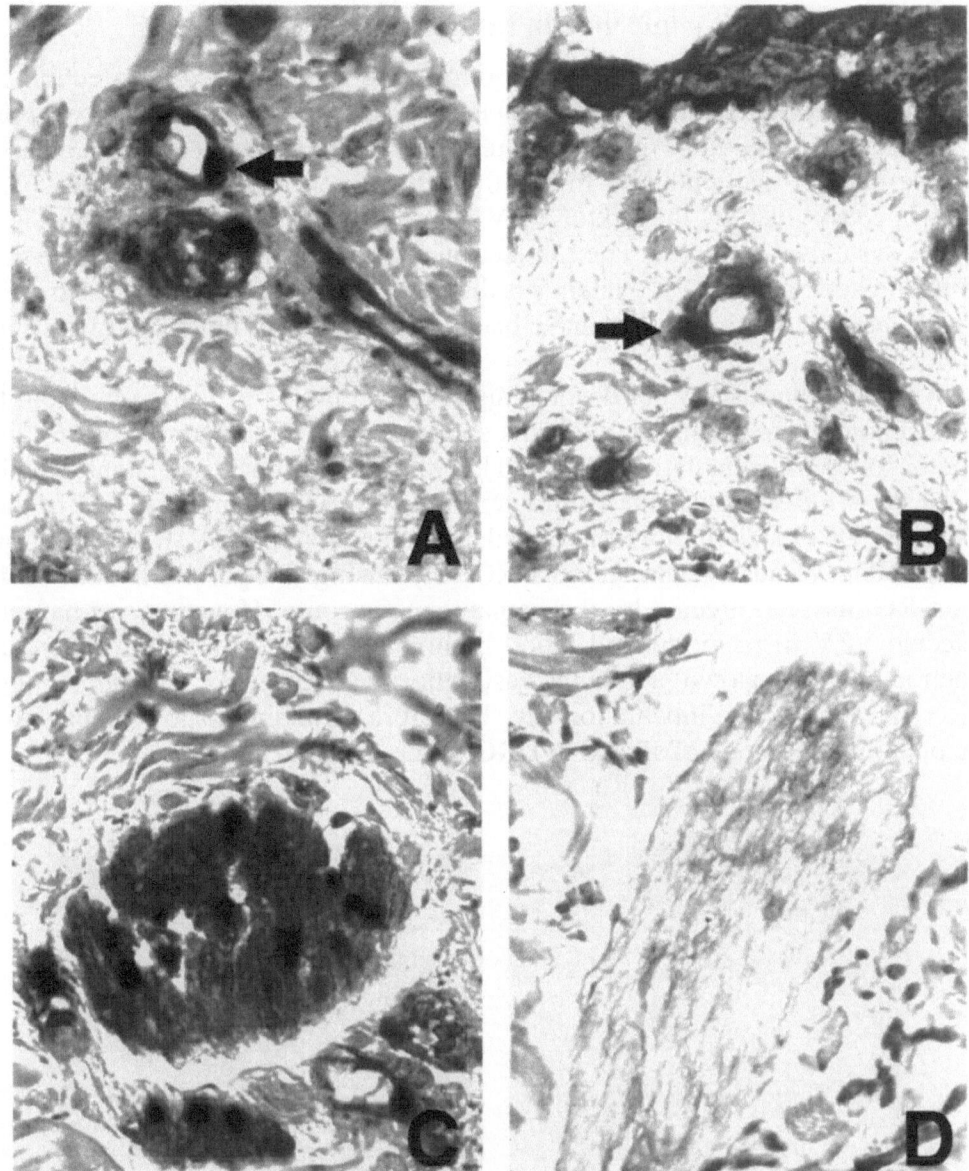

Fig. 2. Immunohistochemical detection of ORF29p and CD43 in skin biopsies. Skin biopsies from a patient with chickenpox (**A, C**) or a patient with zoster (**B, D**) were probed for the presence of ORF29p. The arrows indicate endothelial cells containing ORF29p. **C** and **D** show nerves in which ORF29p was present (**C**) or absent (**D**). The sections are shown at 600× magnification. These panels were reproduced with permission [2]

[17]. The lack of ORF29p from peripheral axons during zoster is consistent with previous findings that this protein localizes to the nucleus during lytic infection and reactivation [13]. These findings raise the question of whether ORF29p has a previously unrecognized role in the initial infection of neurons during chickenpox.

VZV infection in rat dorsal root ganglia

Observations of VZV in human DRG during the initial stages of infection and establishment of virus latency have not been possible because of the lack of appropriate specimens. Human specimens are of limited supply and are not amenable to experimental manipulation. In addition, their usefulness for the study of latency is constrained by the risk of VZV reactivation at the time of death. Tissue culture systems that support latent infection are not available for VZV. The most promising tool to study VZV latency is the rat model [16]. VZV appears to establish latent infection in rats that have been subcutaneously inoculated with virus [1, 8, 16].

Although the rats do not develop signs or symptoms of acute infection, the virus genome is found in satellite cells and in the nuclei of neurons one to three months following inoculation (Fig. 3) [1]. By in situ hybridization, latent infection in the rat is indistinguishable from latent infection in humans (Fig. 3). In most instances, DRG ipsilateral to the inoculation site contain VZV. Rarely, we have found latent VZV in contralateral DRG [1], raising the possibility that in the rat, DRG may be infected by the hematogenous route. However, we have not detected VZV in rat peripheral blood mononuclear cells by polymerase chain reaction (data not shown) and in the vast majority of cases, VZV is found only in DRG ipsilateral to the inoculation site. Therefore, axonal transport appears to be the predominant mode of spread to DRG in the rat model.

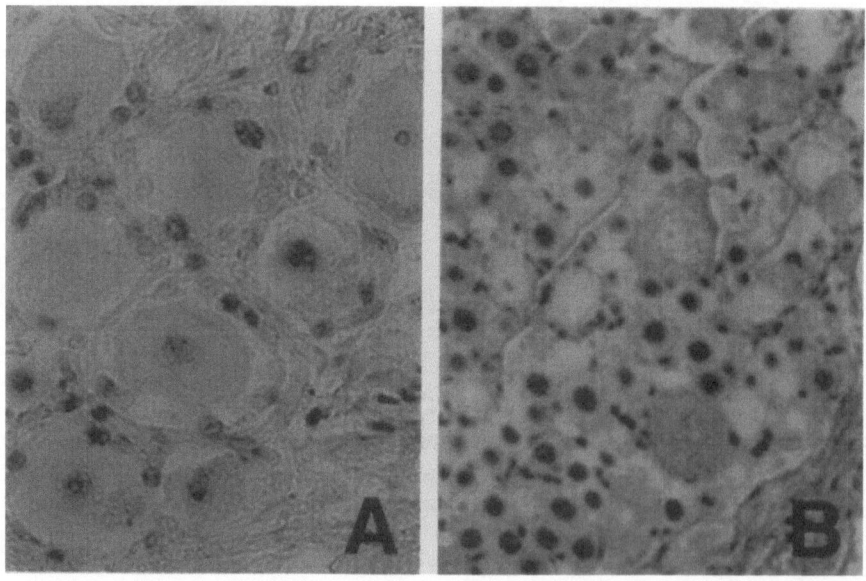

Fig. 3. In situ hybridization detection of VZV DNA in human and rat DRG. DRG obtained from a human without clinical evidence of zoster at the time of death (**A**) and from a rat innoculated with VZV three months prior to sacrifice (**B**) underwent in situ hybridization for VZV VZV DNA is detected in satellite cells and neuronal nuclei in both specimens. The sections are shown at 600× magnification. **B** was reproduced with permission [1]

It remains to be determined if VZV latency in the rat is identical to latency in the human. One VZV protein, IE63p, has been found in rat DRG harboring latent virus [8]. Whether other VZV proteins found in human DRG containing latent virus are present in rat DRG containing latent virus has not been determined. VZV has not been found to spontaneously reactivate in the rat, an important distinction from its behavior in humans that may indicate significant differences between the latency state in the two species. However the lack of spontaneous virus reactivation in rat DRG may be advantageous to investigators studying latency.

Conclusion

Specific immediate early and early proteins accumulate in neurons containing latent VZV and localize to atypical cellular compartments [13]. When VZV reactivates in DRG, the full lytic cycle cascade resumes and these proteins are found in their typical locations [13]. The factors that trigger these changes in protein expression and localization remain unknown. Immunohistochemical analysis of chickenpox and zoster skin lesions may provide clues into the initial establishment and cessation of latency. During primary infection, VZV productively infects endothelial cells, epithelial cells, inflammatory cells bearing monocyte/macrophage markers, and peripheral axons. The presence of ORF29p in peripheral axons during primary infection in puzzling and may suggest that ORF29p is involved in the early stages of neuronal infection. The function that ORF29p performs in peripheral axons during chickenpox does not appear to be required for reactivation since this protein was not found in peripheral axons during zoster. When the virus returns to the skin during zoster, productive infection is again established in epithelial cells, endothelial cells, and inflammatory cells.

Systematic investigations of these events are limited by the lack of appropriate tissues that support latent infection. The rat model may provide a reliable means for studying the establishment and maintenance of latency. Initial studies indicate that latent infection in this model is applicable to latent infection in humans.

Acknowledgements

This work was supported by NIH grants AI-01409 (to PWA) and AI-124021 and by an Irving Scholar Award from the Columbia Presbyterian Medical Center Irving Center for Clinical Research (to PWA).

References

1. Annunziato P, LaRussa P, Lee P, Steinberg S, Lungu O, Gershon A, Silverstein S (1998) Evidence of latent varicella zoster virus in rat dorsal root ganglia. J Infect Dis 178 [Suppl 1]: S48–51
2. Annunziato P, Lungu O, Panagiotidis C, Zhang J, Silvers D, Gershon A, Silverstein S (2000) Varicella-zoster virus proteins in skin lesions: implications for a novel role of ORF29p in chickenpox. J Virol 74: 2005–2010
3. Cohen J, Straus S (1996) Varicella-zoster virus and its replication, vol 2, 3rd ed. Lippincott-Raven, Philadelphia

4. Cohrs RJ, Barbour M, Gilden D (1996) Varicella-zoster virus (VZV) transcription during latency in human ganglia: detection of transcripts mapping to genes 21, 29, 62, and 63 in a cDNA library enriched for VZV RNA. J Virol 70: 2789–2796

5. Cohrs RJ, Barbour MB, Mahlingham R, Wellish M, Gilden D (1995) Varicella-zoster virus (VZV) transcription during latency in human ganglia: prevalence of VZV gene 21 transcripts in latently infected human ganglia. J Virol 69: 2674–2678

6. Cohrs RJ, Srock K, Barbour MB, Owens G, Mahlingham R, Devlin M, Wellish M, Gilden D (1994) Varicella-zoster virus (VZV) transcription during latency in human ganglia: construction of a cDNA library from latently infected human trigeminal ganglia and detection of a VZV transcript. J Virol 68: 7900–7908

7. Croen KD, Ostrove JM, Dragovic LY, Straus SE (1988) Patterns of gene expression and sites of latency in human ganglia are different for varicella-zoster and herpes simplex viruses. Proc Natl Acad Sci USA 85: 9773–9777

8. Debrus S, Sadzot-Delvaux C, Nikkels AF, Piette J, Rentier B (1995) Varicella-zoster virus gene 63 encodes an immediate-early protein that is abundantly expressed during latency. J Virol 69: 3240–3245

9. Gilden D, Rozenman Y, Murray R, Devlin M, Vafai A (1987) Detection of varicella-zoster virus nucleic acid in neurons of normal human thoracic ganglia. Ann Neurol 22: 337–380

10. Kinchington P, Hougland J, Arvin A, Ruyechan W, Hay J (1992) The varicella-zoster virus immediate-early protein IE62 is a major component of virus particles. J Virol 66: 359–366

11. Kinchington PR, Bookey D, Turse SE (1995) The transcriptional regulatory proteins encoded by varicella-zoster virus are open reading frames (ORFs) 4 and 63, but not ORF 61, are associated with purified virus particles. J Virol 69: 4274–4282

12. Lungu O, Annunziato P, Gershon A, Stegatis S, Josefson D, LaRussa P, Silverstein S (1995) Reactivated and latent varicella-zoster virus in human dorsal root ganglia. Proc Natl Acad Sci USA 92: 10980–10984

13. Lungu O, Panagiotidis C, Annunziato P, Gershon A, Silverstein S (1998) Aberrant intracellular localization of varicella-zoster virus regulatory proteins during latency. Proc Natl Acad Sci USA 95: 7080–7085

14. Mahalingham R, Wellish M, Wolf W, Dueland AN, Cohrs R, Vafai A, Gilden D (1990) Latent varicella-zoster viral DNA in human trigeminal and thoracic ganglia. N Engl J Med 323: 627–631

15. Meier JL, Holman RP, Croen KD, Smialek JE, Straus SE (1993) Varicella-zoster virus transcription in human trigeminal ganglia. Virology 193: 193–200

16. Sadzot-Delvaux C, Merville-Louis M-P, Delree P, Marc P, Moonen G, Rentier B (1990) An in vivo model of varicella-zoster virus latent infection of dorsal root ganglia. J Neurosci Res 26: 83–89

17. Schwartz J (1984) Biochemical control mechanisms in synaptic transmissions. In: Kandel E, Schwartz J (eds) Principles of neural science. Elsevier, New York, pp 121–131

Authors' address: Dr. P. W. Annunziato, Department of Pediatrics, 622 West 168th Street, PH West 4-464, New York, NY 10032, U.S.A.

In vitro measurement of human T cell responses to varicella zoster virus antigen

A. R. Hayward

Department of Pediatrics and Immunology, University of Colorado
School of Medicine, Denver, Colorado, U.S.A.

Summary. Means to quantitate cell-mediated immunity are increasingly in demand as modifications to existing vaccines and new vaccines are tested. For immunity to varicella zoster virus, there is over a decade of experience with estimates of responder cell frequency obtained by diluting the number of lymphocytes in antigen-stimulated cultures. This method shows substantial variations between subjects, so populations of 12 or more subjects per group are needed to make comparisons possible. Cytokine-based methods for T lymphocyte responses may prove more sensitive, as may direct antigen-binding methods using tetramers of peptide and histocompatibility antigens – but experience with both is very limited.

Introduction

The development of the varicella zoster virus (VZV) vaccine has stimulated new efforts to measure cell-mediated immune responses to VZV. The aim of these studies has been to define an end point that would serve as evidence for immunity – and, if possible, to define a level of immunity that would indicate that a subject was not at risk for varicella zoster. There is unfortunately no gold standard against which different tests of cell mediated immunity can be judged – and we have no clear understanding of the mechanisms that are important in the immune response to VZV. Quantitating immunity to other pathogens has not often been attempted, although there are some human data for mycobacteria [28] and an impression that the diameter of the tuberculin test for delayed hypersensitivity to PPD is related to immunity [24]. Skin test responses to VZV are measurable in immune subjects and they have the advantage that a response can be measured when it reaches its peak, whether at 24 or 48 h, or later [27]. This is an end point in which Japanese workers have excelled [16] but unfortunately there is no standard, internationally available, VZV antigen for skin test use – and no normative data are available for the experimental antigen preparations.

In vitro tests are at best only poor surrogates for the complex interactions that occur in vivo between virus, antigen presenting cells and responding T lymphocytes. The tests entail making the assumption that blood lymphocytes (the only population that is readily accessible) are representative of the populations of immune cells that act in lymph nodes and tissues. In vitro tests require that the interactions between T lymphocytes and the specialized dendritic cells that normally present antigen in vivo be take place in tissue culture wells in the laboratory. The interactions require live cells, so the blood from which the lymphocytes and monocytes are isolated most be processed within hours of drawing. Different anticoagulants and differences in the handling of the blood are all variables that can affect the outcome.

Lymphocyte stimulation in culture

The most conventional measure of cell-mediated immunity is an antigen stimulation test in which lymphocytes from blood are stimulated by an extract of VZV-infected cells to proliferate. The proliferation requires the synthesis of DNA that is measured by incorporation of tritiated thymidine [29]. Fragments of infected cells are phagocytosed by monocytes in tissue culture and their catabolism within phagolysosomes yields peptides that are presented bound to class II histocompatibility antigens on the monocyte's surface [22]. VZV-immune subjects have VZV antigen-specific CD4 lymphocytes in blood with cell surface receptors complementary to the peptide complexes that have persisted following the primary infection. These cells proliferate in culture and their response is usually measured as uptake of ^{3}H thymidine after 6 days of incubation. Responses in these conventional antigen-stimulated cultures are notoriously difficult to quantitate because there is no linear relationship between the number of responding cells and the amount of radioactive thymidine that is incorporated. Presentation of the antigen by blood cells (monocytes and perhaps to some extent B cells) in culture is likely to be much less efficient than presentation by dendritic cells that occurs in vivo [14]. The antigen itself is subject to batch-to-batch variation. Nevertheless, responses after infection with VZV are greater than before infection and an increase in response has also been observed after immunization. The in-vivo function of VZV-specific CD4 cells is unknown. They can be cytotoxic in tissue culture [11] and can presumably provide help for antibody production [10] and/or modify antigen presenting cells so the latter became capable of providing help for cytotoxic T cells [23].

Some CD8 T cells are activated in cultures of blood lymphocytes stimulated by VZV antigens but it has been hard to confirm that these are indeed VZV-specific. The uncertainty arises because CD8 cells respond to antigenic peptides that are presented in a groove of class I HLA antigens – and the easiest way to determine their specificity is through cytotoxicity tests. A cytotoxicity test requires a target cell that can be killed in culture. The HLA antigens of the target cell should be the same as those on the blood lymphocytes being tested. For human studies a common approach is to grow EBV transformed B cell lines as target cells and to infect these with vaccinia recombinants that express VZV glycoproteins. For the VZV antigens to reach the surface of these target cells (where they can be recognized by cytotoxic T cells) the antigens must first be digested to peptides. These have to be transported to the endoplasmic reticulum where they are incorporated into class I HLA antigens. In vitro, several herpesviruses, including VZV, actively interfere with the transport of peptides or expression of class I HLA antigens on the cell surface [4, 12] and this may limit the role of CD8 T cell responses to the virus. It is difficult to recreate combinations of infected cells with matching histocompatibility antigens in the laboratory, so much less is known of CD8 than CD4 cell responses to VZV.

Studies using VZV antigens purified by immuno-affinity columns suggest that epitopes on the VZV glycoproteins gE and gI stimulate much of the proliferative response and that the IE and nucleoproteins also contribute [1, 2, 25]. Replicate counts of ^{3}H thymidine uptake by cultured lymphocytes vary too much for this method to provide a useful comparison between T cell response to individual epitopes. An alternative but substantially more laborious approach is to measure an individual's response to VZV antigens in cultures in which the number of lymphocytes is reduced by dilution. An assumption implicit in the 'limiting dilution cultures' is that there will be no response from a culture containing no VZV-specific CD4 lymphocytes. Responses by blood lymphocytes from a VZV-immune individual should however be detected when the number of lymphocytes in culture is sufficient to include a cell specific for VZV. If sufficient cultures at each cell concentration are tested then sampling statistics can be used to estimate the fraction of lymphocytes in the population being studied that respond to the antigen – this number is the 'responder cell frequency' or RCF. The estimate of RCF is open to substantial numbers of errors [9] and has to be viewed as an approximation. Alternative approaches such as measuring cytokine production Elispot technology are being explored but they, too, will be subject to sampling errors. It is difficult to control the number of antigen presenting cells per well and the percentage of T cells with the CD4 phenotype will vary between individuals. The increased processing of the blood that would be required to control these variables introduce other variables so that the precision of the test is not generally improved.

Our data using limiting dilution cultures point to a reduction in the number of VZV-responder cells in blood from about 50 per million at age 30 to around 25 per million at age 60 years. The fall continues with further aging unless it is reversed by immunization with VZV vaccine [7] or by an episode of shingles [6]. Curiously, the frequency of T cells responding to VZV antigen in limiting dilution cultures in lower than 50 per million in blood from children under about 12 years of age. Regardless of age, the specificity of most 'responder' cells is confirmed by their ability to kill VZV infected B lymphoblast targets in cytotoxicity tests. Synthesized single VZV epitopes also stimulate lymphocytes but the frequency of responders is substantially lower [8, 15]. The frequency of blood lymphocytes that activates to express the marker, CD69, in VZV-stimulated cultures is about 10 fold higher than that measured by limiting dilution. Some of the cells stimulated to express CD69 on the cell surface by VZV also produce interferon gamma – many of which belong to the CD4 subset [13, 30]. Disparities between the frequency of cells becoming activated and producing cytokines and the frequency detected in limiting dilution cultures are well documented for CD8 cells in mice.

Next-generation tests: tetramer binding by mouse cells

Measurements of mouse T cells with specificity for single epitopes has been revolutionized by the synthesis of labeled peptide-histocompatibility antigen complexes whose binding to the antigen specific receptors on T cells can be directly detected [21]. Studies with synthetic MHC-peptide tetramers in mice show that, before infection, about 2 lymph node cells per million have specificity for an influenza A virus epitope. Infection by an influenza A virus strain bearing the epitope under test stimulates CD8 cells with the corresponding specificity to divide up to 3 times daily until they account for 50% of the cells in the responding lymph node [5]. For the mouse lymphocytic choriomeningitis virus (LCMV) the frequency of tetramer-binding cells reaches 80% in the spleen [3, 19]. Both these figures are 10–100 times greater than the frequency of LCMV-specific cells that is detected by the proliferation-dependent limiting dilution culture assays. The frequency of epitope-specific T cells falls after infection so that (in the case of influenza A virus epitope specific cells) their frequency in spleen in about 1:3000 cells. It is these surviving cells that survive to provide long term specific memory.

The frequency of mouse lymphocytes that binds antigen in tetramer tests approximates to the frequency that makes γ-IFN in culture. It is this similarity in numbers that is the basis for the hope that the frequency of γ-IFN making cells will serve as an end-point for response to infection or immunization. The Elispot method can be applied to human lymphocytes but at this point there are not enough data to assess the strengths and weaknesses in man [17].

Tetramer binding technology was developed in mice because inbreeding yields large numbers of animals with identical histocompatibility antigens. The situation in an outbred and highly polymorhic population such as man is much more complex because individual class I histocompatibility epitopes occur only at low frequency. Data for human CD8 responses is necessarily confined to blood lymphocytes because responding lymph nodes are not generally available. Nevertheless, the frequency of CD8 T cells with specificity for an influenza A matrix peptide in an HLA A0201 subject were generally consistent with the mouse results [20]. Human T cells responding to influenza also make γ-IFN. Estimates of the frequency of responder cells to influenza in man by the antigen binding and intracellular γ-IFN staining approaches are generally similar and are about 10 fold high than the frequency of cells that proliferate in limiting dilution cultures. The basis for these differences is not known. Possible explanations include differences in the long-term survival potential of the responding cell or they may be an artefact of the culture conditions used for limiting dilution.

VZV infects only humans and the principal VZV epitopes that can be recognized by human lymphocytes are not known. Studies of VZV peptide-MHC tetramer binding by human lymphocytes have not yet been reported. The frequency of blood CD4 lymphocytes making γ-IFN in response to VZV is in the 0.2–2% range, one to 2 orders of magnitude greater that the frequency of cells that proliferates in cultures stimulated by VZV antigens. If this figure for cells responding to VZV by making γ-IFN is a true indicator of the frequency of VZV-specific cells, and other herpesviruses stimulate comparable responses, then as many as 10% of blood lymphocytes may have specificity for herpesviruses alone. This seems surprisingly high in relation to the numbers of pathogens that can cause disease in man. Perhaps the frequency of cells that respond to herpesviruses is high because these viruses become latent and antigens produced during latency continue to stimulate memory T cells.

Conclusions

Responses to VZV antigen by human CD4 and CD8 T lymphocytes are detectable after primary infection and after immunization with VZV vaccine. The frequency of the responding cells is low and there are no data indicating the role that these cells play either during latency or during recovery from reactivation of the virus (as in shingles). New technologies such as γ-interferon synthesis and direct antigen binding are being developed and the cytokine techniques should detect sufficient cells for comparisons between responses before and after immunization. Direct antigen binding will be difficult to apply in anything other than very large scale studies because of the low frequency of individual histocompatibility antigens.

References

1. Arvin AM, Kinney-Thomas E, Shriver K, Grose C, Koropchak CM, Scranton E, Wittek AE, Diaz PS (1900) Immunity to varicella zoster virus glycoproteins gp 1 (90/58) and gp III (gp 118) and to a non-glycosylated protein, p170. J Immunol 137: 1346–1351

2. Arvin AM (1996) Immune responses to varicella-zoster virus. Infect Dis Clin North Am 10: 529–570

3. Butz EA, Bevan MJ (1998) Massive expansion of antigen-specific CD8+ T cells during an acute virus infection. Immunity 8: 167–175

4. Cohen JI (1998) Infection of cells with varicella-zoster virus down-regulates surface expression of class I major histocompatibility complex antigens. J Infect Dis 177: 1390–1393

5. Flynn KJ, Belz GT, Altman JD, Ahmed R, Woodland DL, Doherty PC (1998) Virus-specific CD8+ T cells in primary and secondary influenza pneumonia. Immunity 8: 683–691

6. Hayward A, Levin M, Wolf W, Angelova G, Gilden D (1991) Varicella-zoster virus-specific immunity after herpes zoster. J Infect Dis 163: 873–875

7. Hayward AR, Buda K, Jones M, White CJ, Levin MJ (1996) Varicella zoster virus-specific cytotoxicity following secondary immunization with live or killed vaccine. Viral Immun 9: 241–245

8. Hayward AR (1990) T cell response to predicted amphipathic peptides of varicella zoster virus glycoproteins II and IV. J Virol 64: 651–655

9. Hayward AR, Zerbe GO, Levin MJ (1994) Clinical application of responder cell frequency estimates with four years of follow up. J Immunol Methods 170: 27–36

10. Hayward A, Giller R, Levin M (1989) Phenotype, cytotoxic and helper functions of T cells from varicella zoster virus stimulated cultures of human lymphocytes. Viral Immun 2: 175–181

11. Hayward AR, Pontesilli O, Herberger M, Lazslo M, Levin M (1986) Specific lysis of VZV infected B lymphoblasts by human T cells. J Virol 58: 179–184

12. Hill A, Jugovic P, York I, Russ G, Bennink J, Yewdell J, Ploegh H, Johnson D (1995) Herpes simplex virus turns off TAP to evade host immunity. Nature 375: 411–415

13. Jenkins DE, Redman RL, Lam EM, Liu C, Lin I, Arvin AM (1998) Interleukin (IL)-10, IL-12, and interferon-gamma production in primary and memory immune responses to varicella-zoster virus. J Infect Dis 178: 940–948

14. Jenkins DE, Yasukawa LL, Benike CJ, Engleman EG, Arvin AM (1998) Isolation and utilization of human dendritic cells from peripheral blood to assay an in vitro primary immune response to varicella-zoster virus peptides. J Infect Dis 178 [Suppl 1]: 39–42

15. Jenkins DE, Yasukawa LL, Bergen R, Benike C, Engleman EG, Arvin AM (1999) Comparison of primary sensitization of naive human T cells to varicella-zoster virus peptides by dendritic cells in vitro with responses elicited in vivo by varicella vaccination. J Immunol 162: 560–567

16. Kawano S, Terada K, Hiraga Y, Morita T (1996) Immunogenicity of the whole antigen and glycoprotein I of varicella-zoster virus (VZV) and the VZV skin test antigen. Acta Paediatr Jpn 38: 121–123

17. Larsson M, Jin X, Ramratnam B, Ogg GS, Engelmayer J, Demoitie MA, McMichael AJ, Cox WI, Steinman RM, Nixon D, Bhardwaj N (1999) A recombinant vaccinia virus based ELISPOT assay detects high frequencies of Pol-specific CD8 T cells in HIV-1-positive individuals. AIDS 13: 767–777

18. Liu Y, Wehner RH, Zhao M, Nielsen PJ (1997) Distinct costimulatory molecules are required for the induction of effector and memory cytotoxic T lymphocytes. J Exp Med 185: 251–262

19. Murali-Krishna K, Altman JD, Suresh M, Sourdive DJ, Zajac AJ, Miller JD, Slansky J, Ahmed R (1998) Counting antigen-specific CD8 T cells: a reevaluation of bystander activation during viral infection. Immunity 8: 177–187

20. Ogg GS, Jin X, Bonhoeffer S, Dunbar PR, Nowak MA, Monard S, Segal JP, Cao Y, Rowland-Jones SL, Cerundolo V, Hurley A, Markowitz M, Ho DD, Nixon DF, McMichael AJ (1998) Quantitation of HIV-1-specific cytotoxic T lymphocytes and plasma load of viral RNA. Science 279: 2103–2106

21. Ogg GS, McMichael AJ (1999) Quantitation of antigen-specific CD8+ T-cell responses. Immunol Lett 66: 77–80

22. Pontesilli O, Carotenuto P, Levin MJ, Suez DJ, Hayward AR (1987) Processing and presentation of cell-associated varicella zoster antigens by human moncytes. Clin Exp Immunol 70: 127–135

23. Ridge JP, Di Rosa F, Matzinger P (1998) A conditioned dendritic cell can be a temporal bridge between a CD4+ T-helper and a T-killer cell. Nature 393: 474–478

24. Rose DN, Schechter CB, Adler JJ (1995) Interpretation of the tuberculin skin test. J Gen Int Med 10: 635–642

25. Sadzot-Delvaux C, Arvin AM, Rentier B (1998) Varicella-zoster virus IE63, a virion component expressed during latency and acute infection, elicits humoral and cellular immunity. J Infect Dis 178 [Suppl 1]: 43–47

26. Stevenson PG, Doherty PC (1998) Kinetic analysis of the specific host response to a murine gammaherpesvirus. J Virol 72: 943–949

27. Takahashi M, Iketani T, Sasada K, Hara J, Kamiya H, Asano Y, Baba K, Shiraki K (1992) Immunization of the elderly and patients with collagen vascular diseases with

live varicella vaccine and use of varicella skin antigen. J Infect Dis 166 [Suppl 1]: 58–62
28. Van Oers MH, Pinkster J, Zeujlemaker WP (1978) Quantification of antigen-reactive cells among human T lymphocytes. Eur J Immunol 8: 477–484
29. Zaia JA, Leary PL, Levin MJ (1978) Specificity of the blastogenic response of human mononuclear cells to herpesvirus antigens. Infect Immun 20: 646–651
30. Zhang Y, Cosyns M, Levin MJ, Hayward AR (1994) Cytokine production in varicella zoster virus stimulated limiting dilution lymphocyte cultures. Clin Exp Immunol 98: 128–133

Authors' address: Dr. A. R. Hayward, Department of Pediatries and Immunology, University of Colorado School of Medicine, 4200 E 9th Avenue, Denver, CO 80262, U.S.A.

Use of varicella vaccines to prevent herpes zoster in older individuals

M. J. Levin

Department of Pediatrics, University of Colorado School of Medicine,
Denver, Colorado, U.S.A.

Summary. It is likely that the frequency and severity of herpes zoster in older people is the result of an age-related decline in varicella-zoster virus-specific T-cell mediated immunity. Numerous trials of vaccines to boost these responses have demonstrated their safety and immunogenicity. Both live attenuated and inactivated vaccines have been studied. Persistence of booster responses is dose-related, and the half-life of some boosted measures of T-cell mediated immunity exceeds 5 years. Although these trials have been hampered by uncertainty about the critical immune responses to evaluate, the stage is set for a double-blind, placebo-controlled trial of sufficient size to determine efficacy. Such a trial is now underway.

Introduction

It has long been recognized that herpes zoster (HZ) is much more frequent and severe in patients who have defects in T-cell-mediated immunity (CMI) directed against the varicella-zoster virus (VZV), whether by an underlying illness, such as lymphoma, or by immunosuppression with steroids, radiotherapy, or chemotherapy [9, 10, 22, 27]. The singular role of CMI in the pathogenesis of HZ is indicated by the low (normal) incidence of HZ in patients with isolated defects in immunoglobulin synthesis, such as those with inborn errors of immunoglobulin synthesis or with malignancies which depress normal immunoglobulin synthesis (chronic lymphocytic leukemia or multiple myeloma before chemotherapy).

Another well-recognized relationship is the increasing incidence and severity of HZ that accompanies aging [11, 21, 29]. The aging process is characterized by a decline in VZV-specific CMI that correlates closely with HZ morbidity, and it is assumed that these trends are causally related [4, 8, 24, 28, 31]. Interestingly, there is little age-related decline in VZV-specific antibody [4, 12]. These observations naturally suggest that a vaccine capable of boosting VZV-specific CMI in older individuals will prevent (or attenuate) HZ in the vaccinees.

Clinical observations in support of a HZ vaccine

Clinical observations indicate that aging individuals are capable of responding to a VZV antigen challenge and boosting their VZV-specific CMI. For example, the endogenous introduction of VZV antigen with the occurrence of HZ results in an immune response that is adequate to terminate VZV replication within roughly a week of the onset of HZ. Recovery is associated with an increase in VZV-specific CMI [6, 14, 16]. Not only is the immune response to HZ sufficient, it is apparently persistent, since second cases of HZ are very uncommon in people, even those of advanced age [21].

Clinical trials with candidate HZ vaccines

The concept of boosting VZV-specific CMI has been discussed for more than 15 years. At least 7 published and 1 unpublished clinical experiments have been

Table 1. VZV vaccination of older individuals

Author	no.	Age (yr)	Vaccine		Immune test	Duration (yr)
			(PFUx10^3)	antigen U/ml		
Berger[1]	33	55–65	2.7[9]	NA[13]	LPA[14]	<.16
Berger[2]	29	50–65	6–12[9,10]	NA	LPA, Skin Test	<.10
Starr[3]	25	50–74	4.4[11]	6.0	LPA	<.33
Sperber[4]	95	18–49	.28–2.8–28[11] vs nil	.08–8.4	Antibody	1
Takahashi[5]	37	>50	3.0[10]	4.0	Skin Test	1
Berger[6]	200	55–88	3.2–8.5–41.7[12]	0	LPA/RCF[15]/CK[16]	1
Levin[7]	202	60–82	3–6–12[11]	4.0–12.0	RCF/CK/ Antibody	6–8
Hayward[8]	167	55–79	4.0 vs nil	5.2	RCF/CK/ Antibody	3

[1][5]
[2][3]
[3][34]
[4][33]
[5]comm. Takahashi M, Yamanishi K, personal comm
[6][7]
[7][25]
[8][18]; [25a]
[9]Manufacturer: RIT
[10]Received VZV skin test prior to vaccination
[11]Manufacturer: Merck Research Laboratories
[12]Manufacturer: Pasteur Merieux Connaught
[13]Not available
[14]Lymphocyte proliferation assay
[15]Responder cell frequency assay
[16]Induced cytokine assay

undertaken during this interval (Table 1). These experiments vary in important details and have a variety of shortcomings. In the first three studies a generally older population was studied, but the sample size was small and the duration of follow-up was short. The measure of CMI was the lymphocyte proliferation assay (LPA), which has appreciable variability. The first trial of Berger et al. in 33 subjects showed that 17 vaccinees developed a strong VZV-specific LPA response, and 11 developed a weak response [5]. The second Berger trial chose subjects on the basis of a negative VZV skin test, a procedure that is probably immunizing in its own right and would complicate evaluation of the vaccine [3]. In that study 60% of subjects developed a positive LPA after vaccination, but all developed a positive skin test, indicating the possibility that the skin test was a more sensitive indicator of CMI. Starr et al. found that 85% of vaccinees <60 years developed a positive LPA, whereas only 20% of those >70 years did so, raising questions about the age-specific ability to respond to vaccination [34].

The second trial of the Berger group used vaccine doses as high as 12000 pfu. The previous doses had been smaller. Table 1 lists the antigen content of the vaccines, since this component has significant immunogenicity (see below). Sperber et al. vaccinated a large cohort with doses as high as 28000 pfu or an equivalent amount of heat-inactivated VZV [33]. The endpoint of their experiment, specific antibody, may not be relevant to the desired result of vaccination, but the high dose was demonstrated to be safe. The antibody response was the same for the live and inactivated vaccines.

Takahashi administered the current pediatric vaccine, using the skin test to characterize subjects before and at intervals after vaccination. Vaccine-induced immunity was readily demonstrated.

Long-term studies of enhanced VZV-specific CMI

The next three studies are analyzed in more detail, since they were large and/or involve a longer duration of followup. The Berger group used doses as high as 42000 pfu, and applied a less variable – but still imperfect CMI assay – namely, enumeration of circulating VZV-specific memory cells measured with the responder cell frequency assay (RCF) [7]. The next two studies by the HZ study group in Denver have the strength of large sample size, appropriate age of vaccinees, inclusion of a relatively large dose, comparison of an inactivated vaccine, long follow-up, and use of assays for memory cells or lymphocyte cytokine production [18, 25].

Safety

In these three studies more than 480 older people received a live attenuated varicella vaccine; more than 100 received doses in excess of 10,000 pfu. Local reactions, present in less than 35%, consisted of redness, tenderness, and induration. Interestingly, there was no dose effect for reactogenicity and a second booster dose did not produce more severe local reactions. The Berger study included a control group that received pneumococcal vaccine [7]. Varicella vaccine was better tolerated than this control antigen. Approximately 1% of vaccinees in these three studies had a small number of vesicles at the injection site – usually within one week of vaccination. There were a few instances (0.4%) during which vesicles were noted distant from the vaccination site and VZV was detected. However, in general it was unusual to culture virus or detect VZV DNA in post-vaccine lesions. No vaccinees had fever or any serious adverse systemic events. Rare mild systemic symptoms were within the baseline prevalence of minor complaints in this age group. Although the safety profile looks excellent, it is important to realize that the number of vaccinees with more advanced age (e.g. >75 years) who received large doses of vaccine (e.g. >20,000 pfu) is limited. Moreover, it is unknown what would happen if we were to immunize someone with an unrecognized immunosuppressive condition or an elderly person who had not had varicella.

Immunogencity of a shingles vaccine in older people

In the large studies, as well as in the earlier studies, the vaccine induced an increase in VZV-specific memory cells. However, no dose effect was established for the initial response to the vaccine [24]. The immune response included a transient rise in antibody and a rise in γ-IFN that lasted less than one year. Especially noteworthy was a significant increment in RCF. This assay measures primarily CD4+ T cells that express the CD45RO phenotype [19]. This responding population includes cells with MHC class II-restricted cytotoxicity for targets expressing VZV antigens and cells making either γ-interferon or IL-4 [20, 36]. Dose-effects for boosting were absent at 2 years and no age-effects were apparent.

Persistence of booster responses

The study of Levin et al. retained 125 of the original 202 vaccinees 6 to 8 years after vaccination [25]. The half-life of the boost has remained approximately 4.5 years, such that there is still a significant boost apparent at 6 years, albeit appearing to wane (Fig. 1, curve C). While there was no apparent dose effect on the initial response, the persistence studies indicated that RCF was retained longer when a larger dose of virus was administered. This must be accounted for in the final formulation of a shingles vaccine. There are two additional points to make in analyzing these longitudinal data. The first is that the assay is not designed to detect changes in RCF at the extremes of this response; thus, improvement in an already robust RCF can not be readily detected, nor can we accurately detect changes in RCF at or less than 1 VZV-responder cell per 100,000 mononuclear cells. This results in underestimation of the effect of the vaccine. Secondly, although the boost is declining over time with respect to the initial baseline measurement, when one factors in the natural decline in RCF with age (Fig. 1, curve A) it can be seen that a positive vaccine effect persists.

Effects of the vaccine on prevention or attenuation of HZ

There is anecdotal information from approximately 1200 patient-years of followup that the incidence of HZ will not be reduced. There is uncertainty in this

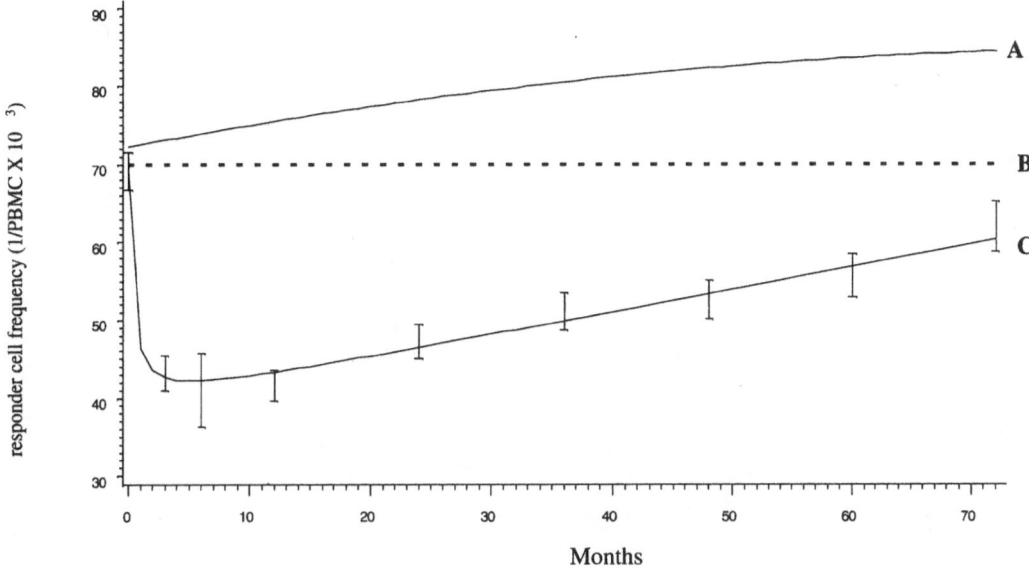

Fig. 1. Mean ($\pm$ SE) of VZV responder cell frequency (RCF) in peripheral blood mononuclear cells (PBMC) from elderly recipients of live attenuated varicella vaccine compared with the expected RCF of a cohort of the same age who had not been vaccinated. Curve A represents expected RCF if the subjects had not been vaccinated. This was calculated from the distribution of RCF values of vaccinees at baseline, using the formula: RCF $= -311.38 + 9.80$ (age) $- 0.06$ (age)^2. Curve B represents the mean baseline RCF of all vaccinees. Curve C represents the RCF of vaccinees at times after vaccination. Frequency at time 0 was significantly less ($P < .05$) than at all other times. Number of observations: 0 months, 188; 3 months, 190, 12 months, 186; 24 months, 177; 36 months, 166; 48 months, 160; 60 months, 153; 72 months, 124 (from [25])

conclusion because we recorded all events in the numerator that were "zoster-like", and when we made most of the observations we did not have PCR available for diagnosis of HZ. However, the clinical impression is that all of these events were unusually mild and pain was rarely significant and never persistent [24, 25]. Thus, it may be that we are converting HZ into a minor illness. We might liken this to "breakthrough HZ" by analogy with the "breakthrough varicella" noted in recipients of the varicella vaccine given to susceptible children.

Experience with an inactivated varicella vaccine in boosting VZV-specific immunity

A comparison of a live varicella vaccine (4000 pfu) and an inactivated vaccine made by heating the live vaccine (5 units of antigen in both) was undertaken in 167 randomly assigned individuals age >60 years ([18]; [25a]). The inactivated vaccine was electrophoretically identical to the live vaccine. As in the previous study, significant antibody and γ-IFN boosts were detected at 3 months, but returned to baseline at one year. The RCF was improved from 1 VZV-responder cell per 65–70,000 mononuclear cells at baseline to 1 VZV-responder cell per 39,000 mononuclear cells at three months post-vaccination (Fig. 2). This was very

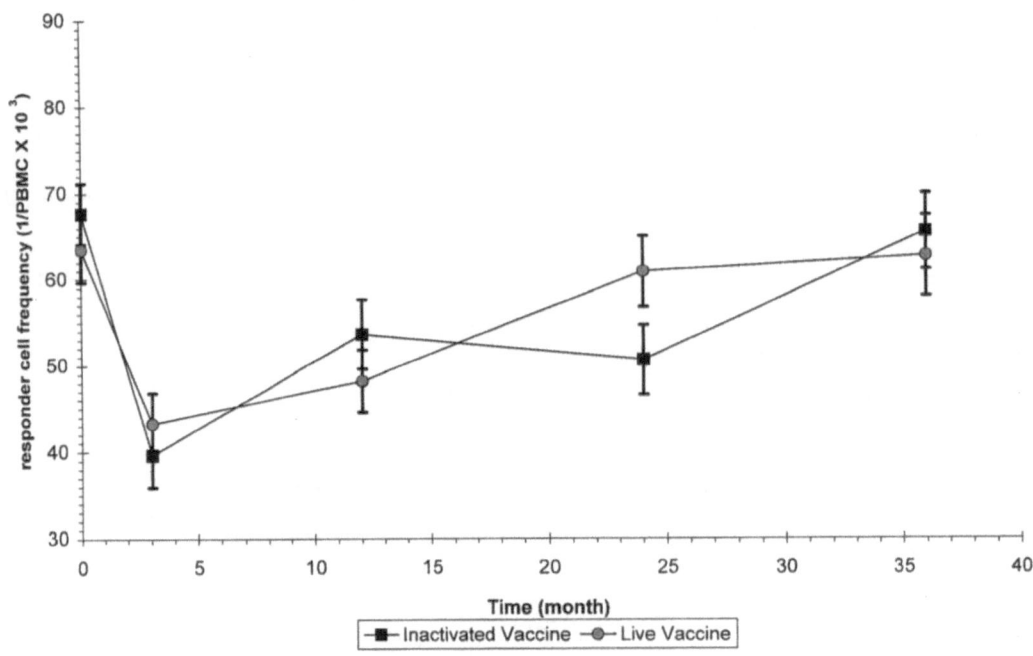

Fig. 2. Mean (± SE) of VZV responder cell frequency (RCF) in peripheral blood mononuclear cells (PBMC) from elderly recipients of live attenuated or inactivated varicella vaccine. Frequency at time 0 was significantly less ($P < .05$) than at 3, 12, and 24 months. There were no significant differences between the two vaccines. Number of observations with live or inactivated vaccine; time 0, 84/81; 3 months, 82/81; 12 months, 71/72; 24 months, 67/66; 36 months, 61/60 (from [25a])

similar to the prior study. The half-life of the response was shorter (23 months); however, the mean dose was also lower for this study and dose influences persistence of response. There was no difference at three years in the boost resulting from either vaccine. Both vaccines increased Class I-restricted cytotoxicity, but the enhancement was greater with the live vaccine [17].

The efficacy of this inactivated vaccine has been piloted in patients with severe defects in VZV-specific CMI [30]. In bone marrow transplant recipients who received inactivated VZV vaccine (4.5 units of antigen; the equivalent of 3000 PFU), given as three consecutive monthly doses starting one month after transplantation, there was no decrease in frequency of HZ (~30%), but the episodes appeared to be attenuated. This mirrors the anecdotal impression derived from studies with live vaccine in the elderly. Thus, an inactivated vaccine might be an effective substitute if the live vaccine causes problems in some groups of elderly subjects.

Evidence that some elderly individuals will fail to respond to active immune boosting

The large studies demonstrated that there is a subset of individuals who have no detectable memory cells prior to vaccination. This subset was significantly

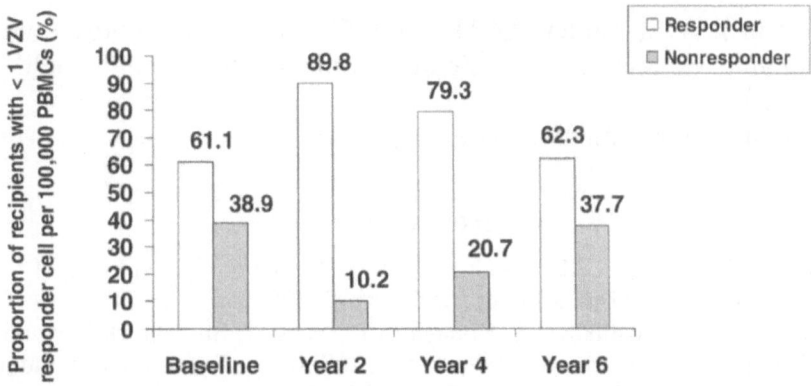

Fig. 3. Presence of ≥ 1 VZV responder cell per 100,000 peripheral blood mononuclear cell (responder) in elderly recipients of live attenuated varicella vaccine at different times after vaccination. Number of observations at each time point is as indicated in Fig. 1 (from [25])

reduced by vaccination, but again grew in size over time (Fig. 3) [25]. The significance of this is uncertain, given the limits of the RCF assay, but the possibility exists that individuals with undetectable memory cells may represent those who are at greatest risk for developing HZ. If that is the case, future trials will need to monitor the proportion of non-responding vaccinees, and this should be a consideration in selecting candidate vaccines.

Conclusions

The stage is set for a sufficiently powered efficacy trial of a shingles vaccine. Such a trial will be placebo-controlled, and will utilize a live attenuated VZV vaccine as potent as the largest doses used in published trials. This vaccine, when evaluated by RCF response in dose-finding pilot experiments, demonstrated a dose-response relationship and was safe when administered to elderly patients, including those with diabetes and chronic obstructive pulmonary disease. The anticipated follow-up period will be at least 3 years in order to accumulate sufficient cases of HZ in the control group. Whatever active immunization strategies are adopted in the future, some epidemiological and environmental factors will be operating simultaneously that might alter the frequency and severity of HZ in future generations. For example, universal immunization is blunting the annual epidemics of varicella in regions with high uptake of the vaccine (Seward J, personal communication). This could impact the natural boosting that may be important for delaying, preventing, or attenuating HZ, whether the individual is immune because of varicella or the vaccine, and could influence the efficacy of a shingles vaccine [1, 2, 13, 23]. Secondly, antiviral therapy of varicella, which occurs in 1–3% of children in the USA [26], could decrease the vigor of VZV-specific immunity after varicella and thereby increase the likelihood of HZ or reduce the response to a shingles vaccine. Finally, there is likely to be less VZV in the ganglia of vaccinees, and this virus may replicate poorly in ganglia, which could result in less, or milder HZ [15, 32]. Conversely, relatively lower VZV-specific immune response after vaccination compared to varicella could have the opposite effect.

The definitive experiment is underway.

References

1. Arvin AA, Koropchak CM, Wittek AE (1993) Immunologic evidence of reinfection with varicella-zoster virus. J Infect Dis 148: 200–205
2. Asano Y, Suga S, Yoshikawa T, Kobayashi I, Yazaki T, Shibata M, Tsuzuki K, Ito S (1994) Experience and reason: Twenty-year follow-up of protective immunity of the Oka strain live varicella vaccine. Pediatrics 94: 524–526
3. Berger R, Amstutz I, Just M, Just V, Luescher D (1985) Booster vaccination of healthy adults with VZV antibody but without a VZV-specific cell-mediated immune response. Antiviral Res [Suppl] 1: 267–271
4. Berger R, Florent G, Just M (1981) Decrease of the lymphoproliferative response to varicella-zoster virus antigen in the aged. Infect Immun 32: 24–27
5. Berger R, Luescher D, Just M (1984) Enhancement of varicella-zoster-specific immune responses in the elderly by boosting with varicella vaccine. J Infect Dis 149: 647–648
6. Berger R, Luescher D, Just M (1985) Restoration of varicella-zoster virus cell-mediated immune response after varicella booster vaccination. Postgrad Med J Postgrad Med J: 61: 143–145
7. Berger R, Trannoy E, Hollander G, Bailleux F, Rudin C, Creusvaux H (1998) A dose-response study of live attenuated varicella-zoster virus (Oka strain) vaccine administered to adults 55 years of age and older. J Infect Dis 178 [Suppl 1]: S99–S103
8. Burke BL, Steele RW, Beard OW (1982) Immune responses to varicella-zoster in the aged. Arch Intern Med: 142: 291–293
9. Derryck A, LaRussa P, Steinberg S, Capasso M, Pitt J, Gershon AA (1998) Varicella and zoster in children with human immunodeficiency virus infection. Ped Infect Dis J 17: 931–933
10. Dolin R, Reichman RC, Mazur MH, Whitely RJ (1978) Herpes zoster-varicella infections in immunosuppressed patients. Ann Intern Med 89: 375–388
11. Donahue JG, Choo PW, Manson JE, Platt R (1995) The incidence of herpes zoster. Arch Intern Med 155: 1605–1609
12. Gershon AA, Steinberg SP (1981) Antibody responses to varicella-zoster virus and the role of antibody in host defense. Am J Med Sci 282: 12–17
13. Gershon AA, Steinberg, NIAID Varicella Vaccine Collaborative Study Group (1990) Live attenuated varicella vaccine: protection in healthy adults compared with leukemic children. J Infect Dis 161: 661–666
14. Hata S (1980) Skin test with varicella-zoster virus antigen on herpes zoster patients. Arch Dermatol Res 268: 65–70
15. Hayakawa Y, Torigoe S, Shiraki K, Yamanishi K, Takahashi M (1984) Biologic and biophysical markers of a live varicella vaccine strain (Oka): identification of clinical isolates from vaccine recipients. J Infect Dis 149: 956–963
16. Hayward A, Levin MJ, Wolf W, Angelova G, Gildend (1991) Varicella-zoster virus-specific immunity after herpes zoster. J Infect Dis 163: 873–875
17. Hayward AR, Buda K, Jones M, White CJ, Levin MJ (1996) VZV specific cytotoxicity following secondary immunization with live or killed vaccine. Clin Immunol 9: 241–245

18. Hayward AR, Buda K, Levin MJ (1994) Immune response to secondary immunization with live or inactivated VZV vaccine in elderly adults. Viral Immunol 7: 31–36

19. Hayward AR, Giller R, Levin MJ (1989) Phenotype, cytotoxic and helper functions of T cells from varicella zoster stimulated cultures of human lymphocytes. Viral Immunol 2: 175–184

20. Hayward AR, Pontesilli O, Herberger, Laszlo M, Levin M (1986) Specific lysis of VZV infected B lymphoblasts by human T cells. J Virol 65: 179–184

21. Hope-Simpson RE (1965) The nature of herpes zoster: a long-term study and a new hypothesis. Proc R Soc Med 58: 9–20

22. Kawasaki H, Takayama J, Ohira M (1996) Herpes zoster infection after bone marrow transplantation in children. J Pediatr 128: 353–356

23. Krause PR, Klinman DM (1995) Efficacy, immunogenicity, safety, and use of live attenuated chickenpox vaccine. J Pediatr 127: 518–525

24. Levin MJ, Hayward AR (1996) Prevention of herpes zoster. Infect Dis Clin North Am 10: 657–675

25. Levin MJ, Barber D, Goldblatt E, Jones M, La Fleur B, Chan C, Stinson D, Zerbe GO, Hayward AR (1998) Use of live attenuated varicella vaccine to boost varicella-specific immune responses in seropositive people 55 years of age and older: Duration of booster effect. J Infect Dis 178 [Suppl 1]: S109–S112

25a. Levine MJ, Ellison MC, Zerbe GO, Barber D, Chan C, Stinson D, Jones M, Hayward AR (2000) Comparison of a live attenuated and an inactivated varicella vaccine to boost the varicella-specific immune response in seropositive people 55 years of age and older. Vaccine 18: 2915–2920

26. Lieu TA, Black SB, Rieser N, Ray P, Lewis EM, Shinefield HR (1994) The cost of childhood chickenpox: parents' perspective. Pediatr Infect Dis J 13: 173–177

27. Ljungman R, Lonnqvist B, Gahrton G, Ringdén O, Sundqvist V-A, Wahren B (1986) Clinical and subclinical reactivations of varicella-zoster virus in immunocompromised patients. J Infect Dis 153: 840–847

28. Miller AE (1980) Selective decline in cellular immune response to varicella-zoster in the elderly. Neurology 30: 582–587

29. Ragozzino MW, Melton LJ III, Kurland LT, Chu CP, Perry HO (1982) Population-based study of herpes zoster and its sequelae. Medicine 61: 310–316

30. Redman RL, Nader S, Zerboni L, Liu C, Wong RM, Brown BW, Arvin AM (1997) Early reconstitution of immunity and decreased severity of herpes zoster in bone marrow transplant recipients immunized with inactivated varicella vaccine. J Infect Dis 176: 578–585

31. Rogers RS III, Tindall JP (1971) Geriatric herpes zoster. J Am Geriar Soc 19: 495–504

32. Somekh E, Levin, MJ (1993) Identification of human dorsal root neurons with wild type varicella virus and the Oka strain varicella vaccine. J Med Virol 40: 241–243

33. Sperber SJ, Smith BV, Hayden FG (1992) Serologic response and reactogenicity to booster immunization of healthy seropositive adults with live or inactivated varicella vaccine. Antiviral Res 17: 214–222

34. Starr SE, Tinklepaugh C, Bocks E, Miller D, Rudenstein M, Plotkin SA (1987) Immunization of healthy seropositive middle aged and elderly adults with varicella-zoster virus (VZV) vaccine. (abstract no. 1237). In Programs and Abstracts of the Twenty-Seventh Interscience Conference on Antimicrobial Agents and Chemotherapy, New York, p 313

35. Yawn BP, Yawn RA, Lydick E (1997) Community impact of childhood varicella infections. J Pediatr 130: 759–765
36. Zhang Y, Cosyns M, Levin MJ, Hayward AR (1994) Cytokine production in varicella zoster virus stimulated limiting dilution lymphocyte cultures. Clin Exp Immunol 98: 128–133

Authors' address: Dr. M. J. Levin, Department of Pediatrics, Pediatric Infectious Diseases, University of Colorado School of Medicine, 4200 East Ninth Avenue, Box C-227, Denver, CO 80262, U.S.A.

Immunization of the elderly to boost immunity against varicella-zoster virus (VZV) as assessed by VZV skin test reaction

M. Takahashi[1], H. Kamiya[2], Y. Asano[3], K. Shiraki[4], K. Baba[5], T. Otsuka[6], T. Hirota[6], and K. Yamanishi[7]

[1] The Research Foundation for Microbial Diseases of Osaka University, Osaka, Japan
[2] National Mie Hospital, Tsu-City, Mie, Japan
[3] Fujita Health University Hospital, Toyoake, Aichi, Japan
[4] Toyama Medical and Pharmaceutical University, Toyama, Japan
[5] Osaka University Hospital, Presently, Baba Pediatric Clinic, Osaka, Japan
[6] Kannonji Institute, The Research Foundation for Microbial Diseases of Osaka University, Kannonji, Kagawa, Japan
[7] Department of Microbiology, Osaka University Medical School, Suita, Osaka, Japan

Summary. The utility of the VZV skin test in detecting individual susceptibility to varicella and zoster was determined. Its specificity particularly with regard to herpes simplex was also established.

The VZV skin test was negative or weakly positive during the early stage of herpes zoster, and strongly positive during recovery from that disease. A small-scale clinical trial to immunize elderly individuals has been performed for the purpose of preventing herpes zoster, and, perhaps, severe post-herpetic neuralgia as well. Sixty individuals ≥50 years old were screened for VZV antibodies by IAHA test and were given a VZV skin test for cell-mediated immunity. All were seropositive, but eight were skin-test negative. Thirty-seven individuals including the eight with negative skin tests were immunized with one dose of varicella vaccine (3.0×10^4 PFU/dose). After 5–7 weeks, the skin test reaction showed increased positivity, with a change in score from (−) to (+, ++) in 7/8 subjects, from (+) to (++, +++) in 3/5 subjects, and from (++) to (+++) in 6/10 subjects. Enhancement of the VZV antibody titer (defined as twofold or greater) was observed in all 15 vaccine recipients with a prevaccination titer of ≤1:16, and in 19 of 24 subjects with a prevaccination titer of ≥1:32.

These results indicate that giving live varicella vaccine with a high viral titer can induce a good boost immunity particularly cell-mediated immunity to VZV in the elderly.

Development of VZV skin test antigen

Cellular immunity to varicella zoster virus (VZV) plays an important role in recovery from or protection against VZV infection. Lymphocyte transformation tests are usually used to test for cell mediated immunity (CMI) to VZV. However, these assays require fresh blood, and are time-consuming. A varicella skin test provided an alternative method to test for CMI to VZV [7, 11]. The varicella skin test was evaluated for its usefulness in determining susceptibility to varicella.

VZV skin test antigen was prepared as follows: VZV-infected human diploid cell cultures were washed three times with phosphate buffered saline (PBS). Cells were collected by trypsinization and washed with PBS by repeated centrifugation at 3,000 **g** for 20 min.

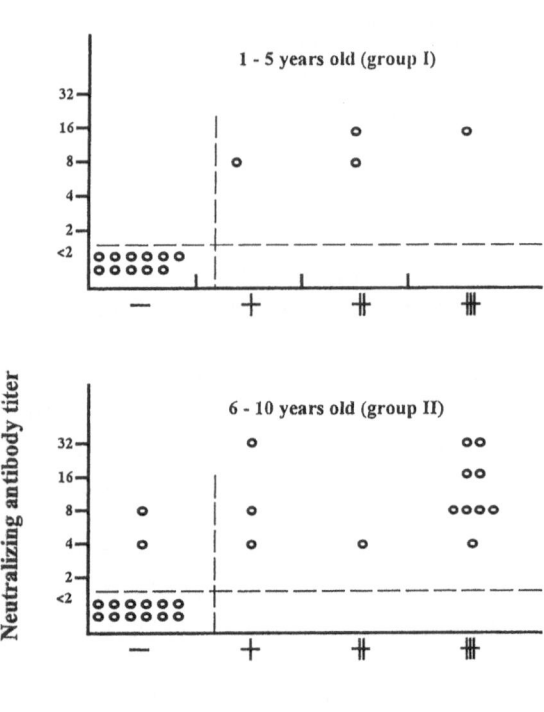
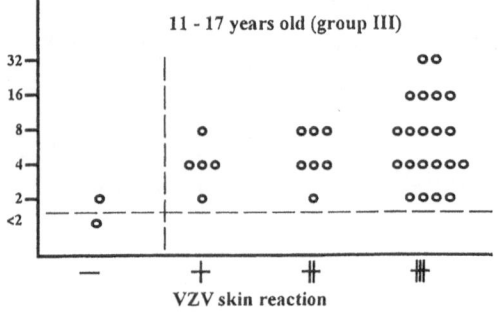

Fig. 1. Correlation between varicella skin test reaction and neutralizing antibody in children and adolescents. The skin test reaction was graded according to the diameter of the erythematous change: $(-) = <5$ mm; $(+) = \leq 5$ mm $-$ <10 mm; $(++) = \geq 10$ mm $-$ <15 mm; $(+++) = \geq 15$ mm (from Kamiya et al. [9])

Clinical application of VZV skin test

The correlation between the varicella skin test reaction in children and neutralizing antibody titer is shown in Fig. 1. Using the criterion of 5 mm of erythema as a positive result, 50 of 53 normal children with a history of varicella and VZV neutralizing antibody had positive skin reactions. In 22 children without a history of varicella and no neutralizing antibody to VZV, skin tests were negative [10].

The skin test was applied prospectively identify susceptible children exposed to a varicella patient in institutional settings. The skin tests were performed immediately after an index case with varicella. Those with negative skin tests were vaccinated, while those with positive skin tests were not vaccinated. None of the children with a positive skin test developed varicella. No spread of varicella was observed except a few cases who were in the incubation period when the vaccine was given [10].

The VZV skin test was again utilized subsequently to rapidly identify varicella susceptibles rapidly when a case of varicella occurred in an institution housing 49 children. Among 20 children with negative skin tests, 17 were vaccinated; the index case and two children with fever were not vaccinated. Only one developed of the 17 children developed varicella which was mild within 2 weeks after vaccination; typical varicella developed in 2 unvaccinated children within 4 weeks after appearance of the index case. None of the 29 exposed children with positive

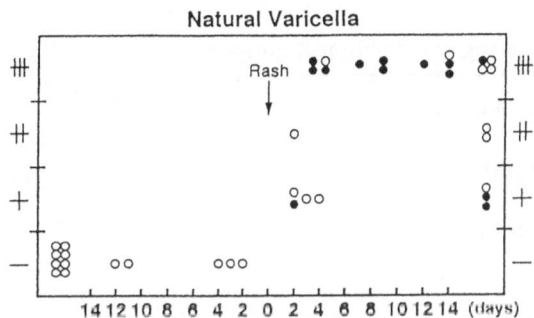

Skin test was performed on 22 children

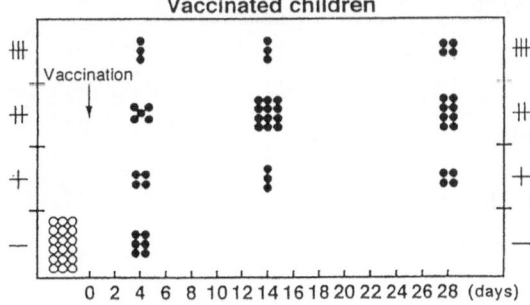

Skin test was performed 4 times in all the 18 vaccinated children

○ indicates the first skin test
● indicates the 2nd. 3rd. 4th skin tests

Fig. 2. Comparison of the time of conversion of varicella skin reaction in children with natural varicella and with vaccination (from Takahashi and Baba [14])

skin test developed varicella [2]. Thus it was proved that the results of VZV skin test accurately reflected the susceptibility of individuals to clinical varicella.

VZV skin test in natural varicella infection and after vaccination

In natural varicella, skin tests were negative in all children before appearance of rash, while positive skin tests were consistently observed after the appearance of rash. Positive skin reactions was detected 4 days after vaccination at the sequential skin test of 4 to 7 days interval (Fig. 2) [16].

Positive lymphocyte transformation was observed a week after vaccination, preceeding the appearance of neutralizing (NT) antibody by 1 to 3 weeks.

An improved VZV skin test antigen

An improved VZV skin test antigen, free from viral particles and cell debris, mainly composed of viral glycoproteins was later prepared by using culture fluid of infected cells as starting material (Table 1). This antigen is far more purified than the original crude VZV skin test antigen, but it still retains antigenicity comparable to that of the crude antigen. The improved skin test antigen was found to be composed of two glycoproteins, gH and gI [13].

Table 1. Flowchart of preparing VZV skin test antigen

Human diploid cells (MRC-5 cells)
↓
Inoculate of VZV (Oka strain)
↓
Wash cells with Earle's solution
↓
Refeed with TCM-199
(without phenol red, with kanamycin 15 mg/ml)
↓
Incubation for 24~48 h at 37 °C
↓
Harvest fluid
↓
Inactivate at 56 °C for 30 min
↓
Centrifuge at 100,000 **g** for 2 h
↓
Harvest the supernatant
↓
Filter and concentrate
↓
Bulk

The relationship of delayed type hypersensitivity (DTH) assessed by the VZV skin test and lymphocyte transformation (LTF)

The relationship of DTH assessed by the VZV skin test and LTF with VZV antigen was investigated in guinea pigs immunized with live VZV vaccine, or with heat-inactivated VZV vaccine. Guinea pigs immunized with live VZV vaccine showed positive DTH and LTF responses to viral antigen as well NT antibody responses, while those immunized with heat-inactivated VZV vaccine virus showed only a NT antibody response of the same degree as that to live vaccine virus [14]. These results demonstrated the reliability of the skin test in assessing CMI to VZV.

Specificity of the skin test with VZV antigen in VZV and with herpes simplex virus (HSV) infection in guinea pigs and in children

Specificity of the skin test with VZV antigen was examined in guinea pigs infected with HSV type 1 or VZV and in children with a history of HSV infection who developed varicella [3]. Infected guinea pigs responded positively only to homologous virus. No cross-reaction between HSV and VZV was detected in the

A   Herpes simplex antibody (IAHA)

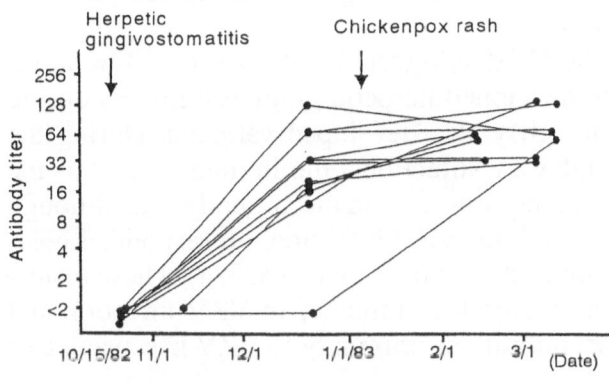

B   Varicella antibody (IAHA)

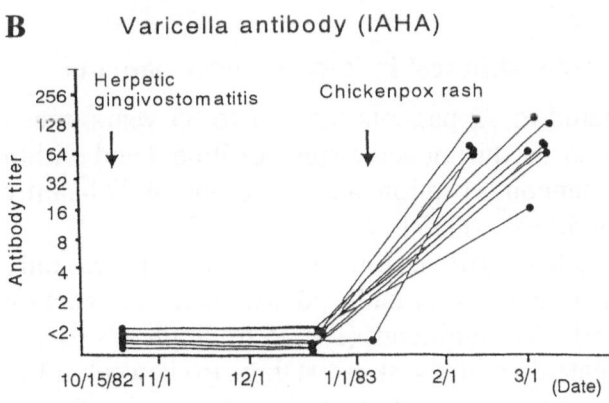

Fig. 3. Immune adherence (IAHA) antibody response to HSV-1 (**A**) and VZV (**B**) in 12 children who were exposed to HSV and then varicella infection. Arrows indicate the onset of outbreak of each epidemic of HSV or VZV. One child had HSV infection just before acquiring chickenpox. All children developed IAHA antibody to HSV and subsequently produced to IAHA antibody to VZV after they had varicella (from Baba et al. [3])

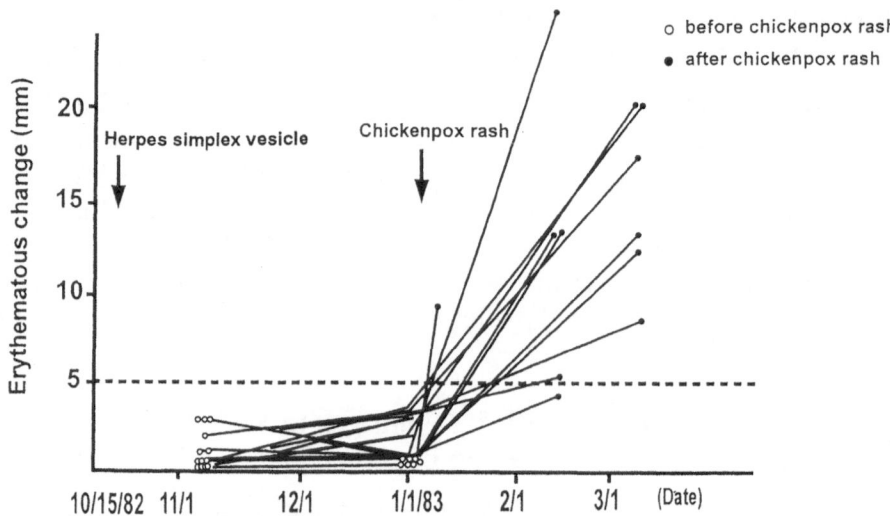

Fig. 4. Cellular immune response assessed by varicella skin test in 12 children who were exposed to HSV and then varicella infection. Skin test reactivity to VZV (measured by erythema size in millimeters) in children remained negative 2–5 months after HSV infection and turned positive during and after development of clinical varicella (from Baba et al. [3])

skin test, or by NT antibody test in infected guinea pigs, indicating that the VZV skin test is specific for VZV infection.

In children, the specificity of the VZV skin test was demonstrated in an institutional setting. Twelve children developed herpetic gingivostomatitis during an HSV epidemic, and subsequently they also developed varicella. During the 2.5 month period between HSV and VZV infections, the immune status of the children to VZV was negative by both the skin test and the antibody test, although antibody to HSV type 1 was detected. After VZV infection, all responded positively by both the skin test and the antibody test to VZV (Figs. 3, 4). These results indicated that the VZV skin test is specific for immunity to VZV infection and would be of value in screening susceptibility or immunity to VZV irrespective of prior HSV infection.

Clinical applications of the VZV skin test in herpes zoster patients

The VZV skin test was administered to 12 patients (age 27 to 85 years, mean 57 years) with ophthalmic herpes zoster during acute stage of illness and within two weeks after the onset of the cutaneous eruption, and in a group of 27 healthy controls (age 18 to 58 years, mean 52 years) [17].

Of the 27 healthy individuals, 25 had positive skin test reactions whereas only one of 12 patients with ophthalmic herpes zoster had a positive skin test reaction (Fig. 5). This difference was statistically significant (P<.05 by Student's t-test). Five of 12 patients were tested repeatedly with the skin test three to six weeks after the onset of eruption. All developed positive test reactions. This study showed that the VZV skin test is a convenient way to diagnose herpes zoster during the

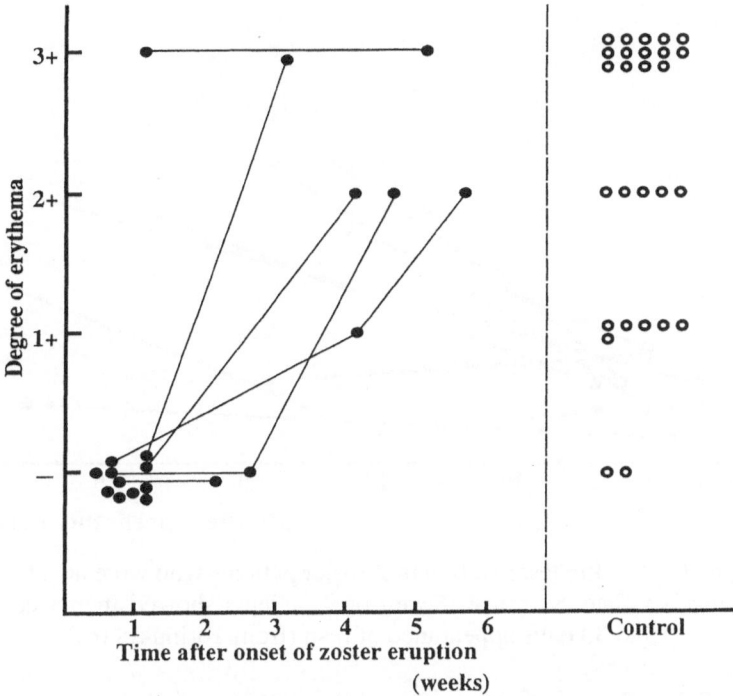

Fig. 5. Result of VZV skin test in 12 patients with ophthalmic zoster (27–85 years mean age 57 years) and control subjects (18 to 58 years, mean age 52 years) (from Tanaka et al. [15])

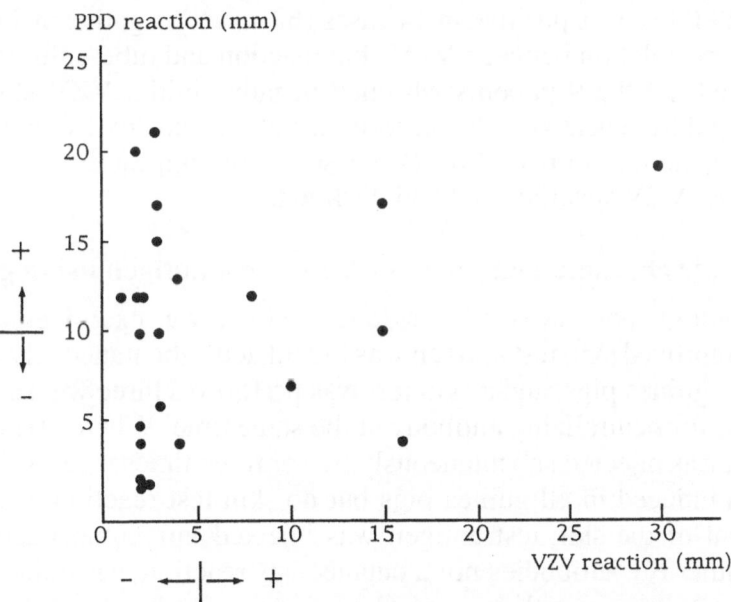

Fig. 6. Relationship between VZV skin test reaction and tuberculin PPD skin test reaction in 23 zoster patients within 7 days after appearance of rash (from Torinuki [16])

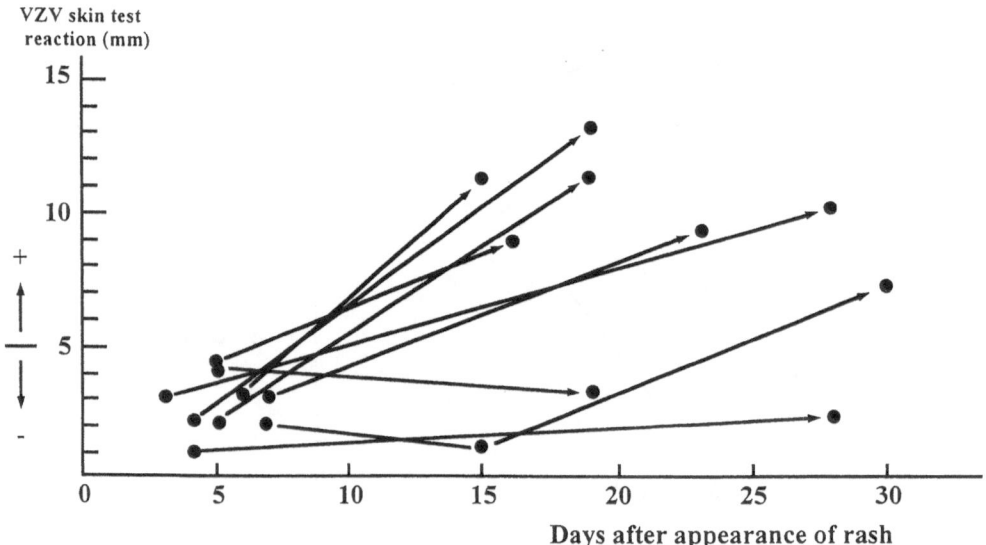

Fig. 7. Change of VZV skin test reaction in 9 zoster patients who were negative in the initial skin test and received later skin test. Seven of 9 patients showed increased skin reaction 15 to 30 days appearance of rash (from Torinuki [16])

acute stage of the disease and suggested that CMI to VZV was impaired at the onset of ophthalmic herpes zoster.

Subsequently, the VZV skin test and tuberculin PPD test were conducted with 23 zoster patients within 7 days after appearance of rash [18]. The VZV skin test was positive in 6 cases (26%) and negative in 17 cases (74%). The tuberculin PPP test was positive in 14 cases (60%) and negative in 9 cases (40%) (Fig. 6). No correlation between VZV skin reaction and tuberculin PPD reaction was observed. Of the 9 patients who had negative initial VZV skin tests, and then examined by repeated skin test seven developed positive skin tests 15 to 30 days after appearance of rash (Fig. 7). These results suggest that zoster tended to develop when VZV-specific CMI had declined.

Immunologic characterization of VZV skin test antigen using guinea pigs

The immunologic potency of VZV skin test antigen was examined using guinea pigs [15]. Improved skin test antigen was inoculated subcutaneously or intracutaneously into guinea pigs, and a skin test was performed three weeks later. Serum was assayed for neutralizing antibody at the same time. When 1.0 ml of the skin test antigen was injected subcutaneously two or three times weekly, NT antibody (4–32) was induced in all guinea pigs but no skin test reaction were observed. When 0.1 ml of the skin test antigen was injected intracutaneously five times weekly, neither NT antibodies nor a cutaneously reaction was induced.

In contrast, live varicella vaccine (1 ml) induced both cutaneous reaction and neutralizing antibody in all guinea pigs. Heat inactivated vaccine also induced both type of immunity in guinea pigs when injected subcutaneously together with Freund's adjuvant.

Table 2. Boosting of immunity against varicella-zoster virus in the elderly by administering live varicella vaccine

Group[a]	Case no.	Age	Sex	Skin test with VZV antigen				Serum antibody titer (IAHA method)	
				Before immunization	After immunization[b]	1 year	2 year	Before immunization[b]	After immunization
I	1 (K.O)	61	m	−	++	++	++	64	128
	2 (N.S)	53	m	−	+	++	●	16	32
	3 (F.S)	54	m	−	+++	+++	+++	16	32
	4 (M.S)	52	f	−	++	++	++	32	64
	5 (C.K)	50	f	−	+++	+++	●	16	64
	6 (S.K)	58	m	−	+++	+++	+++	32	128
	7 (T.K)	59	m	−	++	●	●	4	16
	8 (H.M)	57	m	−	−	−	−	64	64
II	9 (M.S)	53	f	+	+	●	●	64	64
	10 (T.Y)	50	m	+	+	●	●	2	64
	11 (N.O)	55	m	+	+++	●	+++	32	64
	12 (S.I)	53	m	+	+++	●	●	8	16
	13 (N.M)	66	m	+	+++	●	●	64	128
III	14 (K.M)	56	m	++	+++	●	+++	64	128
	15 (I.A)	59	m	++	+	●	●	32	32
	16 (M.A)	59	m	++	+++	●	●	64	≥256
	17 (S.M)	58	m	++	++	●	●	32	128
	18 (T.I)	52	m	++	++	●	++	32	64
	19 (F.M)	54	f	++	+++	●	+++	32	128
	20 (M.F)	54	f	++	+	●	+++	64	128
	21 (K.O)	59	m	++	+++	●	●	16	64
	22 (N.H)	53	m	++	++	●	+++	8	32
	23 (M.Y)	55	m	++	+	●	●	16	128
IV	24 (M.T)	57	m	+++	+++	●	●	64	64
	25 (M.T)	54	m	+++	+++	●	●	64	64
	26 (T.H)	55	m	+++	++	●	+++	32	64
	27 (H.K)	54	m	++	+++	●	+++	16	32
	28 (T.I)	54	m	+++	+++	●	+++	16	32
	29 (N.I)	53	m	+++	++	●	+++	64	128
	30 (K.O)	55	m	+++	+++	●	+++	16	32
	31 (A.S)	52	f	+++	+++	●	+++	16	32
	32 (T.M)	50	m	+++	+++	●	+++	64	128
	33 (Y.H)	65	m	+++	++	●	●	16	32
	34 (H.S)	50	m	+++	++	●	+++	32	≥256
	35 (A.T)	54	m	+++	+++	●	++	32	64
	36 (K.K)	54	m	+++	++	●	+++	128	≥256
	37 (T.U)	52	m	+++	+++	●	●	16	64
	38 (M.N)	60	m	+++	+++	●	●	64	128
	39 (K.T)	54	m	+++	+++	●	●	32	64

[a]Cases were classified as group I, II, III, IV according to the degree of erythematous skin reaction with VZV antigen before immunization. 0∼4 mm (−), 5∼9 mm (+), 10∼14 mm (++), 15 mm ≤ (+++)

[b]After immunization test was done 5∼7 weeks after vaccination

●Not done

These results suggest that immunogenicity of the skin test antigen itself is very low, or nonexistent.

Attempt to boost the immunity of elderly subjects by administering live varicella vaccine

Sixty healthy individuals ≥50 years old were assayed with the VZV skin test for CMI and wit the immune adherence hemagglutinin (IAHA) test for VZV antibodies. Eight of sixty (14%) had negative skin tests, while all were positive for VZV antibodies (Table 2, group I). The degree of the skin test reaction was not necessarily proportional to the antibody titer.

Eight subjects with negative skin tests and 30 with positive skin tests were given live varicella vaccine (3.0×10^4 PFU/dose). The results of the skin tests and antibody assays before and after vaccination are shown in Table 2. The skin test reaction changed from negative to positive in 7/8 subjects, from (+) to (++, +++) in 3/5 subjects, and from (++) to (+++) in 6/10 subjects. Enhancement of the VZV antibody titer was observed in all 15 vaccine recipients with a pre-vaccination titer of ≤1:16, and in 19 of 24 subjects with a prevaccination titer of ≥1:32 (Table 2). The duration of the increased cell-mediated immunity was examined. Four of four subjects examined whose skin test became positive after vaccination maintained the same degree of positivity for at least two years. Fifteen of fifteen subjects whose skin tests were positive at the first screening, and then received vaccination also maintained the same degree of skin test positivity after vaccination for at least two years.

Discussion

We used the VZV skin test and to identify persons susceptibles to VZV, even those with a prior HSV infection [2, 3, 10]. The skin test reaction is a delayed-type hypersensitivity reaction, involving sensitized TH1 lymphocytes which secrete various cytokines, resulting in increased permeability of capillary endothelial cells, leading to erythema or induration.

We showed that the VZV skin test reaction was weak in the early stage of herpes zoster, and increased in its later stages [18]. Enhancement of varicella-specific immune responses occurres in the elderly people who had recovered varicella vaccine [4, 5, 9], and that severe neuralgia was absent in the vaccinated group through six years of follow up; severe neuralgia was seen in control subjects [12]. In all of these studies, CMI was measured with a lymphoproliferative assay [6, 7, 9, 12]. We utilized the VZV skin test reaction for measuring CMI. The stimulation index in the lymphoproliferative reaction is to be comparable to the VZV skin reaction in guinea pigs immunized with VZV [14]. We concluded that the VZV skin test is a convenient tool for monitoring the degree of CMI to VZV in the elderly.

In the present study, it was demonstrated that in most individuals over the age of 50 years who had been given the live varicella vaccine, the skin test reaction and antibody titer was enhanced. Although the duration of immunity should be followed further, these results suggest that enhanced immunity to VZV will lead to prevention of severe herpes zoster. One possible problem in the application of skin test antigen to humans is whether or not booster immune reactions occur after repeated skin tests. However we showed that the immunogenicity of the skin test antigen was detected minimal in guinea pigs [15]. Thus, it seems unlikely that repeated use of skin test antigen itself boosted skin test reactivity in humans. Even if the antigen boosts reactivity in humans, this would not be detrimental as it may boost immunity to VZV.

References

1. Asano Y, Shiraki K, Takahashi M, Nagai H, Ozaki T, Yazaki T (1981) Soluble skin test antigen of varicella-zoster virus prepared from the fluid of infected cultures. J Infect Dis 143: 684–692
2. Baba K, Yabuuchi H, Okuni H, Takahashi M (1978) Studies with live varicella vaccine and inactivated skin test antigen: protective effect of the vaccine and clinical application of the skin test. Pediatrics 61: 550–555
3. Baba K, Shiraki K, Kanesaki T, Yamanishi K, Ogra PL, Yabuuchi H, Takahashi M (1987) Specificity of skin test with varicella-zoster virus antigen in varicella-zoster and herpes simplex virus infections. J Clin Microbiol 25: 2193–2196
4. Berger R, Luescher D, Just M (1984) Enhancement of varicella-zoster-specific immune responses in the elderly by boosting with varicella vaccine. J Infect Dis 149: 647
5. Berger R, Trannoy E, Hollander G, Bailleux F, Rudin C, Creusvaux H (1998) A dose-response study of a live attenuated varicella-zoster virus (Oka strain) vaccine administered to adults 55 years of age and older. J Infect Dis 178 [Suppl 1]: 99–103
6. Chilmonczyk BA, Levin MJ, McDuffy R, Hayward AR (1985) Characterization of the human newborn response to herpesvirus antigen. J Immunol 134: 4184–4188
7. Florman AL, Umland ET, Ballow D, Cushing A, McLaren LC, Gribble TJ, Duncan MH (1985) Evaluation of a skin test for chickenpox. Infect Control 6: 314–316
8. Hayward AR, Herberger M (1987) Lymphocyte responses to varicella-zoster virus in the elderly. J Clin Immunol 7: 174–178
9. Hayward AR, Zebre GO, Levin MJ (1994) Clinical application of responder cell frequency estimates with four years of follow up. J Immunol Methods 170: 27–36
10. Kamiya H, Ihara T, Hattori A, Iwasa T, Sakurai M, Izawa T, Yamada A, Takahashi M (1977) Diagnostic skin test reactions with varicella virus antigen and clinical application of the test. J Infect Dis 136: 784–788
11. LaRussa P, Steinberg SP, Seeman MD, Gershon AA (1985) Determination of immunity of varicella-zoster virus by means of an intradermal skin test. J Infect Dis 152: 869–875
12. Levin MJ, Barber D, Goldblatt E, Jones M, LaFleur B, Chan C, Stinson D, Zerbe GO, Hayward AR (1998) Use of a live attenuated varicella vaccine to boost varicella-specific immune responses in seropositive people 55 years of age and older: duration of booster effect. J Infect Dis 178 [Suppl 1]: S109–112
13. Shiraki K, Takahashi M (1982) Virus particles and glycoprotein excreted from cultured cells infected with varicella-zoster virus (VZV). J Gen Virol 61: 271–275
14. Shiraki K, Yamanishi K, Takahashi M, Dohi Y (1984) Delayed-type hypersensitivity and in vitro lymphocyte response in guinea pigs immunized with a live varicella vaccine. Biken J 27: 19–22
15. Shiraki K, Yamanishi K, Takahashi M (1984) Biologic and immunologic characterization of the soluble skin-test antigen of varicella-zoster virus. J Infect Dis 149: 501–504

16. Takahashi M, Baba K (1984) A live varicella vaccine: its protective effect and immuno-logical aspects of varicella-zoster virus infection. In: De la Maza LM, Peterson E (eds) Medical virology III. Elsevier Biomedical, New York, pp 255–278
17. Tanaka Y, Harino S, Danjo S, Hara J, Yamanishi K, Takahashi M (1984) Skin test with varicella-zoster virus antigen for ophthalmic herpes zoster. Am J Ophthalmol 98: 7–10
18. Torinuki W (1991) Delayed type hypersensitivity skin reaction to both varicella-zoster virus antigen and tuberculin PPD in patients with herpes zoster. Clin Dermatol 6: 381–384 [in Japanese]

Authors' address: M. Takahashi, The Research Foundation for Microbial Diseases of Osaka University, 3-1 Yamadaoka, Suita, Osaka, Japan.

Varicella-zoster virus immunity and prevention: a conference perspective

S. E. Straus

Laboratory of Clinical Investigation, National Institute of Allergy and Infectious Diseases, National Institutes of Health, Bethesda, Maryland, U.S.A.

Summary. This report offers a concise overview of the VZV Conference, highlighting recent developments in the field and speculating on areas of greatest opportunity and need for future work. The goal of eradicating VZV disease will be facilitated by a multifaceted research agenda directed at a fuller comprehension of how the virus replicates, spreads and persists, and how it eludes host immune responses.

Introduction

The international conference whose deliberations are summarized in the foregoing compendium of articles was organized to provide, in part, a research forum. The primary raison d'etre for the conference, however, was the occasion it afforded the participants to celebrate the life and work of Professor Michiaki Takahashi of the Osaka University, on the 25 anniversary of his description of the Oka strain varicella virus vaccine [18, 19], which already has yielded an added measure of health and well-being for millions of children and their parents [9].

The presentations were not meant to convey a comprehensive summary of varicella-zoster virus (VZV) research. Rather, it focused on those topics that together inform the strategies being developed for the ultimate prevention of herpes zoster and its complications. These can be readily divided into subjects relating to VZV epidemiology; the clinical and pathologic features of herpes zoster; new insights into VZV pathogenesis; applications of animal models; a review of humoral and cellular immune responses to VZV; and the use and potential of the Oka vaccine for the prevention of both varicella and herpes zoster. The following synopses of the presentations represent this author's perspectives on the leading issues addressed and of the particular problems yet to be resolved in each of the above fields of inquiry.

Epidemiology

In crafting the first varicella vaccine, Professor Takahashi exploited the then current strategy for development of other live-attenuated vaccines, namely the serial passage of a well characterized viral isolate [18]. Yet, the work was done without any substantive knowledge of the virus, its antigenic stability, or the true manner of its persistence and spread. Insights into these areas required more than the mere clinical description of varicella and analyses of viral seroepidemiology. The best that one could assume at the time is that the strains of VZV that circulate in any seasonal epidemic will be sufficiently similar to those of later epidemics as to confer apparently life-long protection from re-infection. By considering today the antigenic instability of other major viral pathogens – influenza, HIV, and hepatitis C virus – it is indeed fortunate that VZV, and all other herpesviruses apparently have not needed to rely on antigenic shifts to persist in the human host.

This is not to say, however, that the VZV genome is fixed. There is sufficient flexibility in its genome to afford increasingly fine tools for molecular fingerprinting, the ability to track individual VZV strains through the community. From restriction endonuclease analysis to restriction fragment length polymorphism analysis to genomic sequencing, it is evident that small base changes of apparently little or no biologic consequence arise as VZV is passaged serially from person to person [7, 21]. The viruses that circulate in any time or locale represent swarms of VZV variants whose analysis permits us to determine whether a child contracted infection from his grandmother with herpes zoster or from his playmate with varicella. Recent studies revealed that the Oka vaccine is built upon one strain of virus more prevalent in Japan than in Europe and America, and simple PCR-based tools provide ready discrimination among wild-type and vaccine virus. What remains to be understood is the genetic basis for the attenuation of the Oka strain. Initial large-scale sequencing efforts reveal numerous differences among VZV strains, far too many for ready delineation of the most important ones among them (P. Krause and J. I. Cohen, personal communication). It took some three decades to ascertain the basis for attenuation of poliovirus, whose genome size is only a few percent that of VZV, so the challenge is a daunting one, but will be facilitated by the far more sophisticated technologies available today.

Clinical and pathologic features

Remarkably, the compendium of manifestations of VZV infection continues to grow. One would have assumed that all variations of these readily identified exanthematous diseases would by now have been cataloged. Molecular and immunohistochemical tools revealed over the past decade that the immunocompromised individual may become host to a fascinating and perplexing spectrum of infections: disseminated infection without the classic vesicular stigmata of varicella, blinding retinitis due to solely ocular reactivation of VZV, and chronic multifocal encephalitis [2, 4, 17].

Pathogenesis and latency

VZV and herpes simplex virus infections induce classical vesicular lesions that are grossly indistinguishable, yet the propensity of VZV to spread through the epidermal basement membrane into the dermis betrays its proclivity for dissemination. The molecular and biochemical bases for VZV spread remain a challenge that has only partially been met by recent studies of infection in severe combined immunodeficiency (SCID) mice that have been implanted with human tissues [12]. Tedious studies need yet to be conducted to dissect out the VZV genes that facilitate its penetration through vascular walls, its carriage in circulating mononuclear cells during viremic waves, its deposition at distant sites to seed additional crops of lesions, and importantly, its spread and persistence in neural tissues.

VZV latency itself remains a conundrum [15]. While earlier studies differed on the cellular locus of latency, either neurons or non-neuronal cells, recent reports indicate that VZV persists in both types of cells [7, 8]. Latent VZV is associated with continued expression of multiple genes products-five distinct transcripts and their protein products by last count-representing both the immediate early and early kinetic classes [6]. How VZV can persist in both neurons and non-neuronal cells, and accumulate a set of immunogenic proteins within them while eluding host-mediated surveillance and cytotoxic assault needs verification and explication. The expression during latency of several VZV genes begs, moreover, for some understanding of how the replicative cascade is aborted in latency, what sustains the silence of the remaining larger set of viral genes, and what unleashes their expression to incite an episode of zoster.

Immunology

Primary VZV infection is attended by a well-characterized robust and specific humoral antibody response, one which wanes gradually with age. Decades ago, Edgar Hope-Simpson postulated that VZV reactivates subclinically many times in the course of the life of an individual, thereby sustaining humoral immunity, until such time as a clinically evident reactivation event, herpes zoster ensues [5]. Recent studies provide good support for this notion. Subclinical reactivation has been documented in transplant recipients [20], while mathematical modeling of VZV seroepidemiology suggests that only repeated such events could explain the levels at which viral antibodies persist in the population [8a]

It is clear, however, that humoral immunity is not the arbiter of protection against periodic VZV reinfection and reactivation, as these are not characteristic findings in agammaglobulinemic patients. Unfortunately, the nature of the adaptive cellular immune responses that confer this protection are not well understood, despite a vast knowledge of immune responses to many other major human viral pathogens. Efforts need to be mounted to apply existing and evolving tools to define immunogenic epitopes, the targets of cytotoxic T cell responses, and the cascade of cytokines and chemokines that attend immune clearance of primary and reactivation infection with VZV. Once defined, these measures of immune response need to be examined in the context of vaccination, to understand the limitations of the current vaccine and to design better immunoprophylactic and immunotherapeutic reagents.

Animal models

A longstanding obstacle to the analysis of VZV biology, pathogenesis, and immunity has been the lack of small animal models that support infection and emulate the disease spectrum manifested in humans. In the early 1980s, it was shown that guinea pigs can be infected with cell culture-adapted virus and that some infected animals manifest a minor exanthem [13]. The infection was so limited, however, that model never became established as a useful one. In the last decade, it was shown that VZV injected into the flanks of rats accesses the regional dorsal root ganglia and persists there in a form resembling latent infection in humans [14]. More recent studies showed spread of virus to trigeminal ganglia in neonatally-infected rats [1], while the SCID mouse (see above) has been developed to study infection of VZV in selected human tissue implants [12].

None of these systems satisfies the desired properties of a suitable animal model; they each afford only minor glimpses into aspects of pathogenesis. One of the more promising developments has been the serious exploitation of the simian varicella model. The simian virus appears to possess striking genetic relatedness to VZV, induces a severe exanthem, establishes latent infection, and undergoes rare but documented reactivations in Patas monkeys. Over the past two decades, this model has been used primarily to test potential antiviral drugs [16]. Its potential for studies of pathogenesis and latency need to be tapped more fully.

Vaccine

The vaccine that Professor Takahashi wrought is now recommended for universal childhood vaccination in the U.S. It has evinced a remarkable safety record, and longer-term studies suggest that it will sustain protective immunity for at least a decade [3]. Future studies will need to address the question of whether booster doses are required during adolescence or even in adult age. Public health authorities are speculating enthusiastically about prospects for eliminating varicella, but unless truly universal vaccination is achieved, this is not a realistic expectation, given the huge reservoir of latent virus that persists in the billions of us who were once infected.

Two more-feasible ventures should be undertaken, however. The first is to determine whether the existing live, attenuated vaccine strain is able to augment the waning immunity to VZV in senior citizens so as to delay or modify the appearance of herpes zoster [11]. After nearly a decade of planning, a study is now under way to do just that. The Department of Veterans' Affairs in the U.S., in collaboration with Merck and Co. and the National Institute of Allergy and Infectious Disease, is enrolling healthy adults over age 60 into a placebo-controlled vaccine trial (M. Oxman and M. Levin, personal communication), which should be completed in 2005.

The second area of focus must be the creation of a more effective and even safer vaccine. The current vaccine induces weak immunity in adults, it causes varicella in severely compromised recipients, and the vaccine virus establishes latency itself and can reactivate. The time has come to rationally engineer a vaccine or vaccines that do not reactivate and that are more immunogenic. Sufficient knowledge of VZV biology and pathogenesis is being revealed while new classes of vaccines are being fielded in the war on HIV/AIDS [9]. Relevant discoveries need to be exploited to improve our control of VZV and hasten the day when varicella and zoster are eradicated as human diseases.

References

1. Brunell PA, Ren LC, Cohen JI, Straus SE (1999) Viral gene expression in rat trigeminal ganglia following neonatal infection with varicella-zoster virus. J Med Virol 58: 286–290

2. Culbertson WW, Blumenkranz MS, Pepose JS, Stewart JA, Curtin VT (1986) Varicella zoster virus is a cause of the acute retinal necrosis syndrome. Ophthalmology 93: 559–569

3. Garnett GP, Ferguson NM (1996) Predicting the effect of varicella vaccine on subsequent cases of zoster and varicella. Rev Med Virol 6: 151–161

4. Gilden DH, Murray RS, Wellish M, Kleinschmidt-deMasters BK, Vafai A (1988) Chronic progressive varicella-zoster virus encephalitis in an AIDS patient. Neurology 38: 1150–1153

5. Hope-Simpson RE (1975) Postherpetic neuralgia. J R Coll Gen Pract 25: 571–575

6. Lungu O, Annunziato PW, Gershon A, Staugaitis SM, Josefson D, LaRussa P, Silverstein SJ (1995) Reactivated and latent varicella-zoster virus in human dorsal root ganglia. Proc Natl Acad Sci USA 92: 10980–10984

7. Lungu O, Panagiotidis CA, Annunziato PW, Gershon AA, Silverstein SJ (1998) Aberrant intracellular localization of varicella-zoster virus regulatory proteins during latency. Proc Natl Acad Sci USA 95: 7080–7085

8. Kennedy PG, Grinfeld E, Gow JW (1998) Latent varicella-zoster virus is located predominantly in neurons in human trigeminal ganglia. Proc Natl Acad Sci USA 95: 4658–4662

8a. Krause PR, Klinman DM (2000) Varicella vaccination: evidence for frequent reactivation of the vaccine strain in healthy children. Nature Med 6: 451–454

9. Krause PR, Straus SE (1999) Herpesvirus vaccines. Development, controversies, and applications. Infect Dis Clin North Am 13: 61–81

10. LaRussa P, Steinberg S, Arvin A, Dwyer D, Burgess M, Menegus M, Rekrut K, Yamanishi K, Gershon A (1998) Polymerase chain reaction and restriction fragment length polymorphism analysis of varicella-zoster virus isolates from the United States and other parts of the world. J Infect Dis 178 [Suppl 1]: S64–S66

11. Levin MJ, Barber D, Goldblatt E, Jones M, laFleur B, Chan C, Stinson D, Zerbe GO, Hayward AR (1998) Use of a live attenuated varicella vaccine to boost varicella-specific immune responses in seropositive people 55 years of age and older: duration of booster effect. J Infect Dis 178 [Suppl 1]: S109–S112

12. Moffat JF, Zerboni L, Sommer MH, Heineman TC, Cohen JI, Kaneshima H, Arvin AM (1998) The ORF47 and ORF66 putative protein kinases of varicella-zoster virus determine tropism for human T cells and skill in the SCID-hu mouse. Proc Natl Acad Sci USA 95: 11969–11974

13. Myers MG, Stanberry LR, Edmond BJ (1985) Varicella-zoster virus infection of strain 2 guinea pigs. J Infect Dis 151: 106–113

14. Sadzot-Delvaux C, Debrus S, Nikkels A, Piette J, Rentier B (1995) Varicella-zoster virus latency in the adult rat is a useful model for human latent infection. Neurology 45 [Suppl 8]: S18–S20
15. Silverstein S, Straus SE (2000) The pathogenesis of latency and reactivation: the ins and outs of VZV. In: Arvin AM, Gershon AA (eds) Varicella-zoster: basic virology and clinical management. Cambridge University Press, Cambridge
16. Soike KF (1992) Simian varicella virus infection in African and Asian monkeys. The potential for development of antivirals for animal diseases. Ann NY Acad Sci 653: 323–333
17. Stemmer SM, Kinsman K, Tellschow S, Jones RB (1993) Fatal noncutaneous visceral infection with varicella-zoster virus in a patient with lymphoma after autologous bone marrow transplantation. Clin Infect Dis 16: 497–499
18. Takahashi M, Okuno Y, Otsuka T, Osame J, Takamizawa A (1975) Development of a live attenuated varicella vaccine. Biken J 18: 25–33
19. Takahashi M, Otsuka T, Okuna Y, Asano Y, Yazaki T (1974) Live vaccine used to prevent the spread of varicella in children in hospital. Lancet 2: 1288–1290
20. Wilson A, Sharp M, Koropchak CM, Ting SF, Arvin AM (1992) Subclinical varicella-zoster virus viremia, herpes zoster, and T lymphocyte immunity to varicella-zoster viral antigens after bone marrow transplantation. J Infect Dis 165: 119–126
21. Yoshida M, Tamura T, Hiruma M (1999) Analysis of strain variation of R1 repeated structure in varicella-zoster virus DNA by polymerase chain reaction. J Med Virol 58: 76–78

Author's address: Dr. S. E. Straus, Laboratory of Clinical Investigation, National Institute of Allergy and Infectious Diseases, National Institutes of Health, 10 Center Drive, Room 10/11N228, Bethesda, MD 20892-1882, U.S.A.

SpringerJournals

Archives of Virology

Official Journal of the Virology Division of the
International Union of Microbiological Societies

Editor in Chief:	M. H. V. Van Regenmortel, Strasbourg
Editorial Board:	L. Enjuanes, Madrid
	T. Folks, Atlanta, GA
	T. K. Frey, Atlanta, GA
	R. A. Killington, Leeds
	H.-D. Klenk, Marburg
	E. P. Rybicki, Rondebosch
	E. G. Strauss, Pasadena, CA
	M. J. Studdert, Parkville
	M. Tashiro, Tokyo
	A. Vaheri, Helsinki
Virology Division:	M. A. Mayo, Dundee
Special Issues Editors:	C. H. Calisher, Fort Collins, CO
	G. Siegl, St. Gallen

Archives of Virology publishes original contributions from all branches of research on viruses, virus-like agents, and virus infections of humans, animals, plants, insects, and bacteria. Coverage includes the broadest spectrum of topics, from initial descriptions of newly discovered viruses, to studies of virus structure, composition, and genetics, to studies of virus interactions with host cells, host organisms, and host populations.

Multidisciplinary studies are particularly welcome, as are studies employing molecular biologic, molecular genetics, and modern immunologic and epidemiologic approaches. For example, studies on the molecular pathogenesis, pathophysiology, and genetics of virus infections in individual hosts, and studies on the molecular epidemiology of virus infections in populations, are encouraged. Studies involving applied research, such as diagnostic technology development, monoclonal antibody panel development, vaccine development, and antiviral drug development, are also encouraged. However, such studies are often better presented in the context of a specific application or as they bear upon general principles of interest to many virologists. In all cases, it is the quality of the research work, its significance, and its originality which will decide acceptability.

Subscription Information

ISSN 0304-8608 (print), ISSN 1432-8798 (electronic)
2001. Vol. 146 (12 issues). Title No. 705
DM 3910.–, ATS 27493.– plus carriage charges
approx. US $ 2,369.00 including carriage charges

This journal is included in the program:
"LINK – Springer Print Journals Go Electronic"
ISSN (electronic edition): 1432-8798

View table of contents and abstracts online at:
www.springer.at/archvirol

SpringerWienNewYork

A-1201 Wien, Sachsenplatz 4–6, P.O. Box 89, Fax +43.1.330 24 26-62, e-mail: journals@springer.at, Internet: **www.springer.at**
D-69126 Heidelberg, Haberstraße 7, Fax +49.6221.345-229, e-mail: orders@springer.de
USA, Secaucus, NJ 07096-2485, P.O. Box 2485, Fax +1.201.348-4505, e-mail: orders@springer-ny.com
Eastern Book Service, Japan, Tokyo 113, 3–13, Hongo 3-chome, Bunkyo-ku, Fax +81.3.38 18 08 64, e-mail: orders@svt-ebs.co.jp

SpringerNews

Martin H. Groschup,
Hans Kretzschmar (eds.)

Prion Diseases

Diagnosis and Pathogenesis

2000. IX, 290 pages. 89 partly coloured figures.
Hardcover DM 250,–, öS 1750,–
(recommended retail price)
(Special edition of "Archives of Virology", Supplement 16)
(Softcover edition of Supplement 16 only available
for subscribers of "Archives of Virology")
ISBN 3-211-83530-X

A comprehensive understanding of the biology of prion diseases is crucial for risk assessment as regards both humans and animals. To further the communication of the current knowledge and the exchange of diagnostic technologies in the scientific community, a symposium on the 'Characterization and Diagnosis of Prion Diseases' was held from September 23-25, 1999 in Tübingen, Germany. The overwhelming interest and the participation of more than 500 scientists from 26 countries made this symposium the largest meeting ever held in the field.

In this Supplementum to Archives of Virology twenty-six invited speakers comprehensively present their data on the pathogenesis of prion diseases in humans and animals, on molecular mechanisms involved in the transmissibility across species barriers, on animal and in-vitro models currently available for the detection and quantification of infectivity and on the characterization of prion strains. This selection attempts to reflect the current state of the art, but cannot possibly represent the whole spectrum of research in the prion field.

SpringerLifeSciences

 **SpringerWienNewYork**

A-1201 Wien, Sachsenplatz 4–6, P.O. Box 89, Fax +43.1.330 24 26, e-mail: books@springer.at, Internet: **www.springer.at**
D-69126 Heidelberg, Haberstraße 7, Fax +49.6221.345-229, e-mail: orders@springer.de
USA, Secaucus, NJ 07096-2485, P.O. Box 2485, Fax +1.201.348-4505, e-mail: orders@springer-ny.com
Eastern Book Service, Japan, Tokyo 113, 3–13, Hongo 3-chome, Bunkyo-ku, Fax +81.3.38 18 08 64, e-mail: orders@svt-ebs.co.jp

SpringerLifeSciences

Charles H. Calisher, M. C. Horzinek (eds.)

100 Years of Virology

The Birth and Growth of a Discipline

1999. VII, 220 pages. 75 figures.
Hardcover DM 248,–, öS 1736,–
(recommended retail price)
(Special edition of "Archives of Virology, Supplement 15")
(Softcover edition of Suppl. 15 only available
for subscribers to "Archives of Virology")
ISBN 3-211-83384-6

One hundred years ago, when Martinus W. Beijerinck in Delft and Friedrich Loeffler on Riems Island discovered a new class of infectious agents in plants and animals, a new discipline was born.
This book, a compilation of papers written by well-recognized scientists, gives an impression of the early days, the pioneer period and the current state of virology. Recent developments and future perspectives of this discipline are sketched against a historic background.

With contributions by A. Alcami, D. Baulcombe, F. Brown, L. W. Enquist, H. Feldmann, A. Garcia-Sastre, D. Griffiths, M. C. Horzinek, A. van Kammen, H.-D. Klenk, F. A. Murphy, T. Muster, R. O'Neill, P. Palese, C. Patience, R. Rott, H.- P. Schmiedebach, S. Schneider-Schaulies, G. L. Smith, J. A. Symons, Y. Takeuchi, V. ter Meulen, P. J. W. Venables, V. E. Volchkov, V. A. Volchkova, R. A. Weiss, W. Wittmann, H. Zheng

Please visit our website: **www.springer.at**

SpringerWienNewYork

A-1201 Wien, Sachsenplatz 4–6, P.O.Box 89, Fax +43.1.330 24 26, e-mail: books@springer.at, Internet: **www.springer.at**
D-69126 Heidelberg, Haberstraße 7, Fax +49.6221.345-229, e-mail: orders@springer.de
USA, Secaucus, NJ 07096-2485, P.O. Box 2485, Fax +1.201.348-4505, e-mail: orders@springer-ny.com
Eastern Book Service, Japan, Tokyo 113, 3–13, Hongo 3-chome, Bunkyo-ku, Fax +81.3.38 18 08 64, e-mail: orders@svt-ebs.co.jp

SpringerLifeSciences

Philip S. Mellor et al. (eds.)

African Horse Sickness

1998. VIII, 342 pages.
86 partly coloured figures.
Hardcover DM 290,–, öS 2030,–
(recommended retail price)
(Special edition of "Archives of Virology, Supplement 14")
(Softcover edition of Suppl. 14 only available
for subscribers to "Archives of Virology")
ISBN 3-211-83133-9

African horse sickness virus is a double-stranded RNA virus which causes a non-contagious, infectious arthropod-borne disease of equines and occasionally dogs. Nine distinct, internationally recognised serotypes of the virus have so far been identified. This book is based upon the findings of two programmes funded by the European Commission. It will be of value not only to the specialist research workers but also to veterinary workers dealing with control and to legislators seeking to promote safe international movement of equines. The topics covered include state-of-the-art discussions on diagnostics, vaccines, molecular biology, vector studies, and epidemiology.

"... This splendid and highly entertaining book gives a comprehensive account of the result obtained from those studies. Its contents are organised along disciplinary lines and are divided into sections dealing with epidemiology, entomology, molecular biology, vaccines and diagnosis ..."

The Veterinary Record

Please visit our website: **www.springer.at**

 **SpringerWienNewYork**

A-1201 Wien, Sachsenplatz 4–6, P.O.Box 89, Fax +43.1.330 24 26, e-mail: books@springer.at, Internet: www.springer.at
D-69126 Heidelberg, Haberstraße 7, Fax +49.6221.345-229, e-mail: orders@springer.de
USA, Secaucus, NJ 07096-2485, P.O. Box 2485, Fax +1.201.348-4505, e-mail: orders@springer-ny.com
Eastern Book Service, Japan, Tokyo 113, 3–13, Hongo 3-chome, Bunkyo-ku, Fax +81.3.38 18 08 64, e-mail: orders@svt-ebs.co.jp